T0318379

PARTICULATES MATTER

Emerging Issues in Analytical Chemistry

Series Editor
Brian F. Thomas

Co-published by Elsevier and RTI Press, the *Emerging Issues in Analytical Chemistry* series highlights contemporary challenges in health, environmental, and forensic sciences being addressed by novel analytical chemistry approaches, methods, or instrumentation. Each volume is available as an e-book, on Elsevier's ScienceDirect, and in print. Series editor Dr. Brian F. Thomas continuously identifies volume authors and topics; areas of current interest include identification of tobacco product content prompted by regulations of the Family Tobacco Control Act, constituents and use characteristics of e-cigarettes and vaporizers, analysis of the synthetic cannabinoids and cathinones proliferating on the illicit market, medication compliance and prescription pain killer use and diversion, and environmental exposure to chemicals such as phthalates, endocrine disrupters, and flame retardants. Novel analytical methods and approaches are also highlighted, such as ultraperformance convergence chromatography, ion mobility, in silico chemoinformatics, and metallomics. By highlighting analytical innovations and new information, this series advances our understanding of chemicals, exposures, and societal consequences.

2016 *The Analytical Chemistry of Cannabis: Quality Assessment, Assurance, and Regulation of Medicinal Marijuana and Cannabinoid Preparations*
Brian F. Thomas and Mahmoud ElSohly

2016 *Exercise, Sport, and Bioanalytical Chemistry: Principles and Practice*
Anthony C. Hackney

2016 *Analytical Chemistry for Assessing Medication Adherence*
Sangeeta Tanna and Graham Lawson

2017 *Sustainable Shale Oil and Gas: Analytical Chemistry, Geochemistry, and Biochemistry Methods*
Vikram Rao and Rob Knight

2017 *Analytical Assessment of e-Cigarettes: From Contents to Chemical and Particle Exposure Profiles*
Konstantinos E. Farsalinos, I. Gene Gillman, Stephen S. Hecht, Riccardo Polosa, and Jonathan Thornburg

2018 *Doping, Performance-Enhancing Drugs, and Hormones in Sport: Mechanisms of Action and Methods of Detection*
Anthony C. Hackney

2018 *Inhaled Pharmaceutical Product Development Perspectives: Challenges and Opportunities*
Anthony J. Hickey

2018 *Detection of Drugs and Their Metabolites in Oral Fluid*
Robert White

2020 *Smokeless Tobacco Products*
Wallace Pickworth

2020 *Analytical Methods for Biomass Characterization and Conversion*
David Dayton

PARTICULATES MATTER

Impact, Measurement, and Remediation of Airborne Pollutants

VIKRAM RAO

Research Triangle Energy Consortium, Research Triangle Park, NC, United States

WILLIAM VIZUETE

University of North Carolina, Chapel Hill, NC, United States

Elsevier
Radarweg 29, PO Box 211, 1000 AE Amsterdam, Netherlands
The Boulevard, Langford Lane, Kidlington, Oxford OX5 1GB, United Kingdom
50 Hampshire Street, 5th Floor, Cambridge, MA 02139, United States

Published in cooperation with RTI Press at RTI International, an independent, nonprofit research institute that provides research, development, and technical services to government and commercial clients worldwide (www.rti.org). RTI Press is RTI's open-access, peer-reviewed publishing channel. RTI International is a trade name of Research Triangle Institute.

Library of Congress Cataloging-in-Publication Data
A catalog record for this book is available from the Library of Congress

British Library Cataloguing-in-Publication Data
A catalogue record for this book is available from the British Library

ISBN: 978-0-12-816904-9

For information on all Elsevier publications
visit our website at https://www.elsevier.com/books-and-journals

Publisher: Susan Dennis
Acquisitions Editor: Kathryn Eryilmaz
Editorial Project Manager: Devlin Person
Production Project Manager: Bharatwaj Varatharajan
Cover Designer: Matthew Limbert
Cover Art: Dayle G. Johnson

Typeset by SPi Global, India

Dedication

To my father, Mannige Vaman Rao, mathematician. Had he continued beyond his MS to a PhD at King's College London, as exhorted by his advisor (c. 1938), his 1939 wedding in India would most likely have been later. Whereas I might still exist, I would not be the same person.[a] This book may well have not been written.

[a] Parfit, Derek (1984). *Reasons and Persons*. Oxford University Press.

Contents

List of contributors

Christina E. Murata, PhD

Managing Director, Deloitte Consulting, Washington, DC, United States

Vikram Rao, PhD

Executive Director, Research Triangle Energy Consortium, Research Triangle Park, NC, United States

William Vizuete, PhD

Associate Professor, Department of Environmental Sciences and Engineering, Gillings School of Global Public Health, University of North Carolina at Chapel Hill, Chapel Hill, NC, United States

Sarah Zelasky, MS

Environmental Risk Assessment Specialist, United States Environmental Protection Agency, Research Triangle Park, NC, United States

Preface

A colleague once related to me (V.R.) that her mother would tell her how inside each of our bodies were molecules from the cosmos and how we are literally built from the stuff of stars. And, the story would continue, with each breath we share the same air as the giants of history such as Caesar, Napoleon, and Churchill. From stardust to Shakespeare, the individual connection with the cosmos and the past appealed to her inner dreamer. And although this tale from childhood is true, it brings with it a darker reality. In many places in the world, each breath brings with it an insidious stowaway, one that slowly deteriorates your health, a villain that robs you of quality and quantity of life. This nemesis is particulate matter (PM).

Airborne PM is arguably the single greatest public health scourge, accounting for over 6.5 million deaths per annum worldwide. Ninety-eight percent of people living in urban areas in low- and middle-income countries live with ambient PM above World Health Organization guidelines. Indoor air pollution from the burning of woody biomass for cooking accounts for about 4.3 million deaths annually. This level of mortality is staggering. To provide some context: annual deaths attributable to air pollution are six times and four times those from malaria and HIV-AIDS, respectively. The mortality and morbidity impact has deep economic consequences for those countries that are most affected. A recent World Bank report estimated more than USD 5 trillion in lost economic productivity. This is equivalent to the combined annual gross domestic products (GDPs) of Canada, Mexico, and India. In some regions of the world, the air pollution challenge retards GDP growth by over 7% each year. The human costs from air pollution are far-reaching and merit a coordinated and sustained mitigation effort.

The PM world is divided by size into PM_{10} and $PM_{2.5}$, those particles with diameters less than 10 μm and 2.5 μm, respectively. The physiological responses in humans are distinctly different for each size, even accounting for the fact that the chemical compositions vary. $PM_{2.5}$ is generally recognized as by far the more dangerous. Interestingly, although most of the particles are combustion-related inorganic carbon, the virulence is considerably greater when they acquire an organic surface. They do so by reactions with species in the atmosphere, some of which, such as volatile organic compounds, can also be anthropogenic. This book covers the cellular and molecular biology implications of human health. We also explore the burgeoning field of ultrafine particles, that is, nanomaterials, that are smaller than 100 nm and are designated as $PM_{0.1}$. The occurrence falls into two categories: ultrafine particles incidental to combustion processes that also produce larger particles and deliberate production for a purpose. Such purpose could be industrial or therapeutic. In the latter category, advantage is taken of the fact that these particles penetrate into the cells, allowing them to carry a diagnostic or therapeutic payload. This ability to cross cell boundaries is also the reason airborne $PM_{0.1}$ is particularly dangerous. Evidence is presented to support the hypothesis that the $PM_{0.1}$ component of $PM_{2.5}$ may be responsible for most of the toxic impact on human organs. Conclusive proof would dramatically affect policy, which currently is mass based; the United States regulates PM to be under 12 $\mu g/m^3$. Ultrafine particles are so small that the mass associated with them is tiny. Declaring victory by eliminating 95% of the mass could be premature; most of the ultrafine particles could still be present, and little would have been achieved on the mortality front. A comprehensive policy would necessarily include limits on *particle count*, not just mass.

A first principles approach to PM amelioration dictates that one must be able to quantify the problem. In some cases, this is easier said than done. For example, statistics of indoor PM_{10} and $PM_{2.5}$ concentrations are hard to come by, particularly in areas of the world, where populations lack electricity and are forced to use wood or kerosene for cooking and lighting. One feature of this book is to evaluate the state of the art in measuring and evaluating PM in all settings and to identify advances in that space. Over time, one could expect more precise estimates in all areas of human endeavor. Because of the cost and logistics of distributing measurement devices, much of the insight regarding spatial distribution of species is acquired through satellite imagery. Additional granularity is obtained by distributing discrete measuring stations at intervals and inserting the data into the models associated with the satellite data.

Another dimension of the book is an overview of the principal anthropogenic sources of airborne particulates: combustion of coal for power and wood burning for cooking and heat, and emissions from internal combustion engines, both stationary and mobile. We discuss these sources and posit the notion that amelioration can most usefully be accomplished through preventing the emissions in the first place. This is unlike actions for other airborne pollutants such as sulfur compounds, for which the principal mitigation method is capture. We discuss engineered solutions in each of the three areas, some proven and others still experimental. A key feature of the proposed mitigations is an emphasis on practical pathways for implementation. Almost by definition, this means we go beyond mere technological wizardry for solutions to this quintessential multidimensional challenge. We offer a consideration of the economic, political, and social elements that must also be mustered for sustained improvements. The chapter on electric vehicles underlines these considerations. The principal allure is the zero tailpipe emissions. But electricity production worldwide is still dominated by coal-fired plants. This is especially the case in India and China, where electric vehicles could make the greatest inroads, in part because fleets are being supplemented, not replaced. The analogy of the mushrooming of cell phones in these countries is valid, despite being cloyingly overused. The economic hurdle of battery cost is close to being traversed. Finally, the most important reason for electric vehicles may well be that the well-to-wheel energy usage is nearly a third of that in gasoline engines. One simply uses much less energy per kilometer travelled.

One chapter is devoted in part to an extraordinary recent discovery, that PM deposits on solar panels impair their efficiency. When present together with dust, the reduction is nearly double that with dust alone. Absent intervention, the net impact on an important renewable could be substantial. An irony lies in the fact that solar power is regarded as an alternative to the burning of $PM_{2.5}$-emitting coal for electricity. This is especially the case in India and China that are burdened with serious levels of PM, particularly in urban settings. We describe research under way to illuminate the issue.

Although the situation is dire, it need not remain so. Fundamentally, this book is grounded in the optimism that air quality challenges can be successfully addressed. One need look no further than our success with lead poisoning. We removed the primary sources, lead in gasoline and paint. In the case of paint, the titanium oxide substitute was even a superior product. But in both cases, entire industries, and the buying public, were needed to embrace the solution. Here, too, the solutions will be some combination of technology, public policy, and societal acceptance.

The nemesis may yet meet its match.

Acknowledgments

Many contributed references, illustrations, reviews of portions of the text, and other manners of support. A few did a lot more than that, and their contributions are identified in greater detail. All comments that follow are those of the lead author (V.R.).

Sarah Zelasky contributed about as much as anybody could, short of being a coauthor. She has been involved from the beginning in 2017, when she worked with me as a summer research assistant, as a fresh graduate of the Gillings School of Global Public Health at UNC, Chapel Hill. Her support continued sporadically through her graduate studies at Harvard. Sarah performed literature searches for most of the chapters, produced illustrations, proposed outlines for several chapters, and wrote the first draft of Chapter 5, *Health effects*.

Christina Murata was an early coauthor of this book. When she left RTI International, her new commitments did not permit continued coauthorship. Her contributions were in the early deliberations on the identification of the target audience and the content to appeal to the target. Her detailed contributions are sprinkled in the Preface and the early chapters, especially Chapter 4, *The importance of being small.*

Prakash Doraiswamy (RTI International) provided guidance in satellite coverage and associated modeling in Chapter 6, *Detection and evaluation of airborne particles*. Ryan Chartier (RTI International) provided figures and reviewed portions of the text in Chapter 8, *Wood fires: domesticated.* Michael Valerino (Duke University) made available a prepublication manuscript (since published) used in Chapter 13, *Clean coal and dirty solar panels*.

PART ONE

PM and ozone fundamentals

This section is about mechanisms. It is largely concerned with how particulate matter (PM) and ozone form and their impact on the human condition. Over 80% of PM is sourced from the combustion of coal, wood, and hydrocarbons. The mechanisms of formation are essentially the same in all three cases, with differences that influence the relative toxicities.

Ozone is addressed in part because it plays a role in the toxicity of PM. It also has a direct impact on health. The other reason for discussing ozone is that its formation in the troposphere is by the reaction of NO_x with volatile organic chemicals, both of which are products of combustion that also yields PM.

The section also addresses the question of why we should care about PM and ozone. Ozone is the principal constituent of smog, together with its precursor chemicals and PM. Ozone is directly responsible for illness, especially in children. PM is responsible for mortality and morbidity. Whereas a host of organs are affected, the principal entry point is the respiratory system. We touch on the physiology of the action.

Finally, we describe the journey taken from early realizations of airborne particulates being harmful to the modern era of regulations.

CHAPTER ONE

Why particulates matter

Early days: Recognition of the issue

Air pollution affecting visibility has been noted for millennia. Impact on health has been recognized since the 12th century.[1] In the 17th century, the Italian physician Ramazzini studied the connection to occupational respiratory illnesses in 52 job settings. Interestingly, the parameter he used was particle count, not mass, estimated with an early version of the optical microscope.[2] But systematic investigation of the effects of particulate matter (PM) inhalation was not done until the mid-20th century. Fig. 1.1 shows some of the earliest quantitative estimations of deposition by anatomical locus. The data are from studies conducted in the late 1980s.

The principal finding in the figure is that roughly 50% of 10 μm particles penetrate the mouth. That figure is roughly 60% for 2.5 μm particles penetrating the nose. Studies such as this were influential in the selection of PM_{10} and $PM_{2.5}$ as standards for regulation. Ultrafine particles (less than 0.1 μm) have a high proportion being deposited in the lung. Deposition in the lung certainly is dominated by particles smaller than 2.5 μm when compared with larger particles. This confirms the decision to focus on that size as opposed to 10 μm. In fact, in the early going during the 1970s, 15 μm was established as the upper end of the coarse range.[3] However, the studies upon which the estimates are premised were criticized for the unrealistic means for delivery of the aerosols to the human subjects.[4] Eventually, the US Environmental Protection Agency (EPA) settled on PM_{10} as a regulatory measure.

Studies such as those that generated the data in Fig. 1.1 emphasized the role of the fine fraction 2.5 μm as a cutoff was being considered; 1 μm also got a lot of support, because, as is evident in Fig. 1.1 that size is closer to the 50% absorption point in the nose than is the 2.5-μm size. Epidemiological studies using $PM_{2.5}$ inlets[5,6] were influential in the EPA settling on $PM_{2.5}$ as the fine particle standard.[7] Some of the decision appears premised upon the

Particulates Matter
https://doi.org/10.1016/B978-0-12-816904-9.00016-7

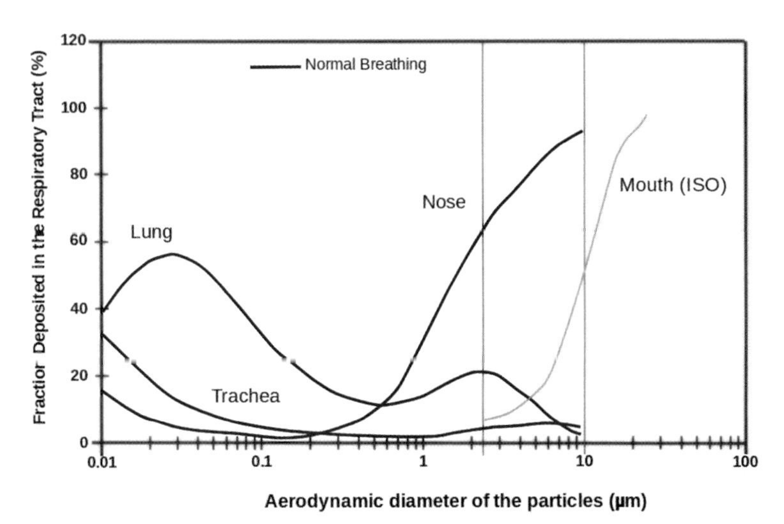

Fig. 1.1 Inhalation and deposition of particulate matter in the human respiratory system. *(Modified from Chow JC. Critical review: measurement methods to determine compliance with ambient air quality standards for suspended particles.* J Air Waste Manage Assoc. *1995;45:320–382.)*

availability of $PM_{2.5}$ inlets and the resultant standardization of aerosol delivery for epidemiology studies.

Progress of regulations

National ambient air quality standards (NAAQS) for PM and other air contaminants were first set in the United States and were emulated in some fashion by other countries. These standards cover a lot of other pollutants beyond PM. For PM, they specify averaging times of the measurement and the concentration level for each period. The US 1970 Clean Air Act calls for a reevaluation of the NAAQS standards every 5 years.

Table 1.1 shows evolution of the standards in the United States. The PM_{10} standard was vacated in 2006. The 24-h standard was maintained at 150 μg/m^3 but is omitted from the table. The annual average $PM_{2.5}$ standard was tightened in 2013 to 12 μg/m^3; the 24-h standard was left at 35 μg/m^3.

The World Health Organization (WHO) promulgates guidelines that are used by jurisdictions to set standards but have no legal authority. The current WHO values for $PM_{2.5}$ are 25 and 10 for the 24-h and annual averages, respectively.

Table 1.1 Evolution of particulate matter standards in the United States.

Year of standard	Species	Twenty four hours' average[a] ($\mu g/m^3$)	Annual average[b] ($\mu g/m^3$)
1987	PM_{10}	150	50
1997	$PM_{2.5}$	65	15
2006	$PM_{2.5}$	35	15
2013	$PM_{2.5}$	35	12

[a]Ninety-eighth percentile averaged over 3 years.
[b]Arithmetic mean averaged over 3 years.

Current state of the particulate matter problem

The WHO guidelines are routinely exceeded in most large cities worldwide. Table 1.2 lists the top 10 polluted cities. The WHO conducts studies spanning 5 years; the last one was for 2008–13 covering 67 countries, including 795 cities. Worldwide pollution increased by 8%, despite improvements in some cities. WHO's ability to conduct these studies is strongly dependent on all jurisdictions using the same instruments and standards. This underlines the need for consistent global standards. A refrain through this book is that the ultrafine fraction is likely disproportionately toxic when compared with the rest of $PM_{2.5}$. The need for consistency over decades is an important reason to not attempt to supplant the standard with one for $PM_{0.1}$. This caution holds even though the relative virulence of $PM_{2.5}$ when compared with PM_{10} pretty much rendered the latter passé.

Table 1.2 Cities with the highest annual PM averages.

City/country	$PM_{2.5}$ ($\mu g/m^3$)	PM_{10}($\mu g/m^3$)
Peshawar, Pakistan	111	540
Rawalpindi, Pakistan	107	448
Gwalior, India	176	329
Hamad Town, Bahrain	66	318
Allahabad, India	170	317
Delhi, India	143	292
Greater Cairo, Egypt	117	284
Pasakha, Bhutan	150	275
Raipur, India	144	268
Varanasi, India	146	260

Data from Borten L. *MSN news*; 2020. https://www.msn.com/en-au/news/photos/cities-with-the-dirtiest-air-in-the-world/ss-BBSM0A9?viewall=true#image=1. Retrieved 29 July 2020, original data WHO 2018 assessment.

Also, in Chapter 13, we describe the negative impact of anthropogenic PM on solar panel efficiency. Interestingly, the most deleterious sizes are larger than 500 nm, and $PM_{0.1}$ has no material impact.

Over half the cities on the most polluted list are either in India or Pakistan. Of note is that Beijing, often held up as the standard for bad air pollution, is not on the list, almost certainly because of the measures taken. Similarly, Delhi dropped down, but not off, in part because of the Indian Supreme Court action that caused all public transport to switch to compressed natural gas fuel.

Health effects

Over 6 million premature deaths per annum are attributed to PM. Eighty-seven percent of these deaths occur in low- and middle-income countries, which represent 82% of the population of the world. That is more than six times the deaths attributed to malaria. Notable is that mortality estimates are not actual counts but are based on models that in turn obtain the data for the parameters from a variety of sources, including epidemiological studies. This is one reason that estimates vary. The figure noted above is from the WHO.[8] Chapter 9 describes the methodology for estimating mortality attributed to wildfires.

Pollutants damage human health by two routes. Waterborne pollutants such as heavy metals, aromatic compounds, and, most recently entering our consciousness, microplastics enter the body through the gastrointestinal tract. Airborne pollutants such as PM infiltrate through inhalation. Fig. 1.2 is a schematic of the human respiratory tract. Size and shape of particles largely determine the retention in the extrathoracic region. As already discussed, regulations relied on the relative disposition of particles in mouth versus nasal breathing. Chemical and biological properties play a part in interaction with cells and tissue, although size plays the prominent role. The larger fraction of fine particles is captured in the nose, rather than the mouth. Breathing rate can affect the degree of absorption in the different parts of the respiratory tract, but for simplicity of presentation, we selected the normal breathing statistics in Fig. 1.1. The smallest particles, especially those in the ultrafine region, make their way down to the bronchioles and alveoli.

Fig. 1.3 details a study of absorption in the entire tract, particularly in the alveolar region, which is shown in the exploded view in Fig. 1.2. Fine and coarse particles are deposited mostly in the extra-alveolar region. Ultrafine

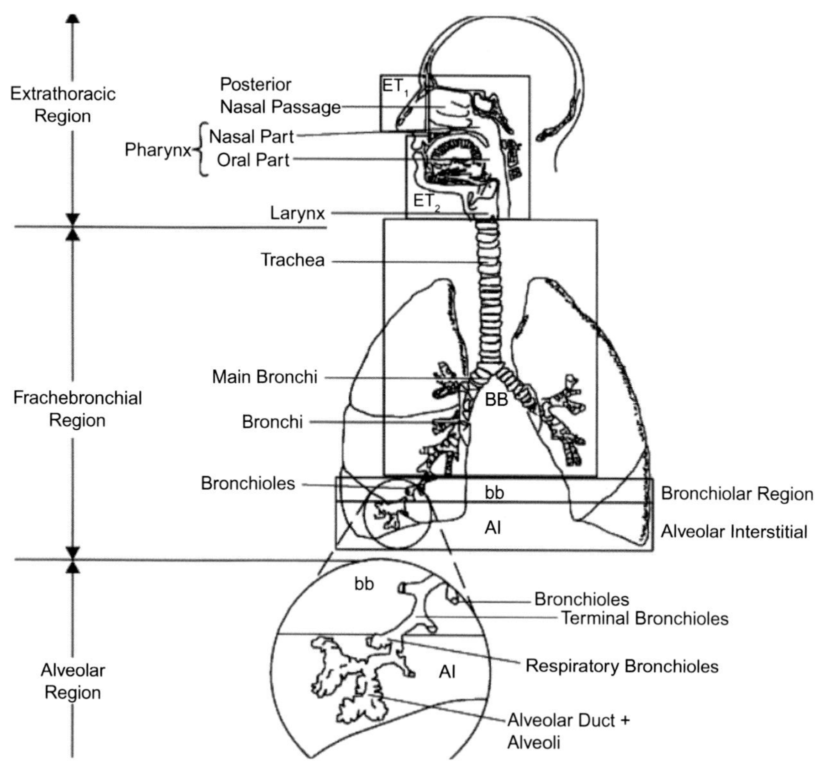

Fig. 1.2 The human respiratory tract, including details of the alveolar region.[9]

particles make their way into the periphery of the lung. Alveolar deposition is in the range of 8–80 nm and peaks at 20 nm. This is consistent with observations later in the book that the ultrafine range is proving the more toxic portion of $PM_{2.5}$.

As the first point of entry for inhaled species, the respiratory tract is the most likely entrée for PM, no matter the final organ affected. The list of organs and associated diseases is long and getting longer with time and better investigative methods.

Cardiovascular disease is strongly correlated with elevated levels of fine PM. Two mechanisms are identified: change in heart rhythm and systemic inflammation. Quickening of the pulse and less variability in the heart rate have been observed. Less variability could make the heart prone to irregular heartbeat that is a known precursor for cardiac arrest. Inflammation in various parts of the body can promote blood clotting and constriction of blood vessels. Inflammation is known to be a factor in hardening of the arteries

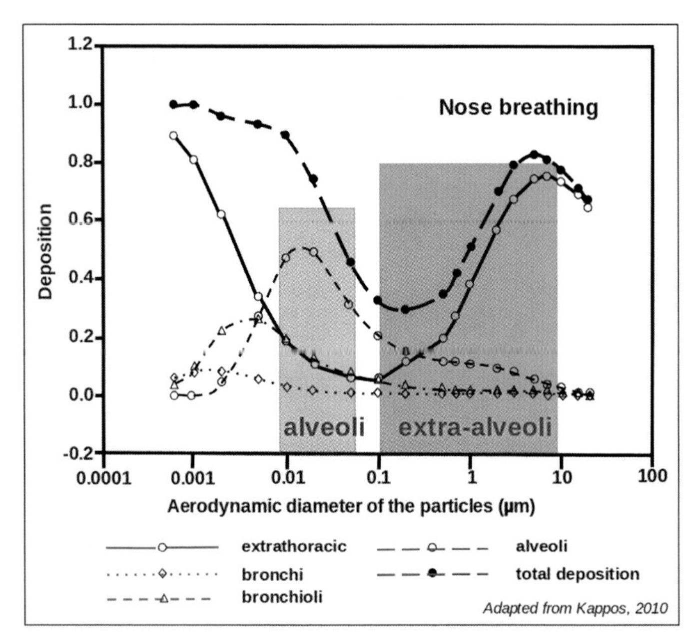

Fig. 1.3 Distribution of PM deposition in the respiratory tract by particle size (logarithmic scale) during light breathing in a healthy person. *(Modified from Kappos A. Health risks of urban airborne particles.* Environmental Science and Engineering (Subseries: Environmental Science)*; 2011. https://doi.org/10.1007/978-3-642-12278-1_27 (web archive link).)*

(atherosclerosis). All these factors can have other causes as well, such as smoking.

A recent epidemiological study indicates that each 10 μg/m^3 increase in ambient PM is associated with a 16% increase in mortality from heart disease and a 14% increase in mortality from stroke.[10] This sort of report increases pressure on the EPA to reduce the current annual average standard of 12 μg/m^3. The probability of this happening is discussed in Chapter 14.

Impact of elevated particulate matter on severe acute respiratory syndrome and Covid-19

The Covid 19 pandemic commenced in late 2019 and spread widely in early 2020. At this writing (June), the end is not in sight. Epidemiologic studies have attempted to link elevated PM with the severity of the disease. Severe acute respiratory syndrome (SARS) predated Covid-19. Its causal agent is also a coronavirus, with features shared by SARS-CoV-2, the agent associated with Covid 19. One such feature is that they both enter cells using the angiotensin-converting enzyme 2 receptor (see the close-up later). SARS did not have the sustained global impact of the current disease, but the mechanisms investigated in SARS have proven instructive. A study of SARS mortality in China found a direct correlation with the air pollution index (API), which is roughly the same as the EPA's air quality index. Both measure concentrations of PM, sulfur dioxide, nitrogen dioxide, carbon monoxide, and ozone; and both report the species with the highest concentration in the index. In the China study, this species was PM_{10}. It would include the $PM_{2.5}$ fraction but that was not split out, likely because the relative importance of the smaller particle size range had not yet been recognized in 2002, or even more likely because the air monitoring stations deployed at the time estimated only PM_{10}.

Fig. 1.4 shows that the relationship between API and SARS mortality is reasonably well behaved but the correlation coefficient is not very high. The striking finding is that patients diagnosed with SARS are twice as likely to die in high API areas than those in low API areas.[11] The authors also studied the effect of long-term exposure. Whereas the effects were greater in the more polluted areas, the data were inconsistent, the trend demonstrating

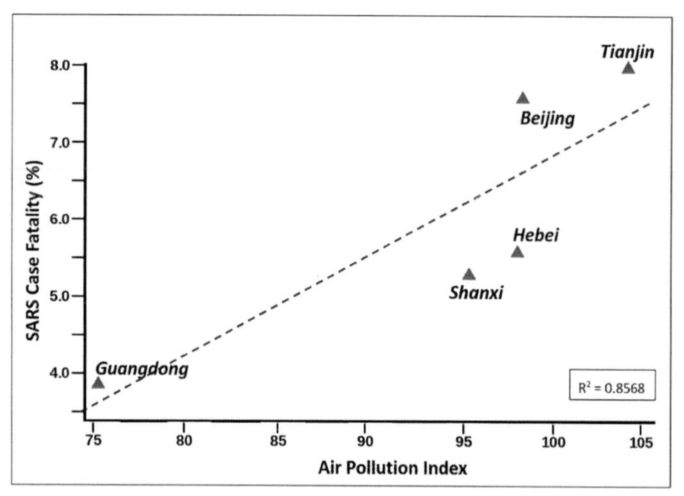

Fig. 1.4 Severe acute respiratory syndrome fatality as related to short-term exposure to ambient PM. Fatality as percentage of severe acute respiratory syndrome diagnosed patients as a function of API was measured in April and May 2003. PM_{10} was designated as the principal pollutant.[11] *(From Cui Y, Zhang ZF, Froines J, Zhao J, Wang H, Yu SZ, Detels R. Air pollution and case fatality of SARS in the People's Republic of China: an ecologic study.* Environ Health. *2003;2:15.)*

a maximum at an intermediate level of pollution. The curious point is that API under 100 is considered acceptable for public health. All the data are under 105, ordinarily a nominally safe zone. This finding appears to bolster the school of thought that no level of pollution is safe. More likely is that confounding factors were not fully considered. Still, the main finding of Cui et al. was given credence by being cited in a news story[12] as generally consistent with the finding by Wu et al. of a correlation between long-term PM exposure and mortality associated with Covid-19.[13] They cited the SARS study as support for their observations.

At the time of this writing, the Wu et al. study, which was from the School of Public Health at Harvard University, was available only as a preprint submitted to the *New England Journal of Medicine* and not yet peer reviewed. It used data from 3087 counties in the United States. Death counts were up to April 20, 2020. The conclusion is that long-term exposure to $PM_{2.5}$ increases the vulnerability to a more severe outcome from contracting Covid-19. The authors accounted for 20 confounding factors in coming to their conclusion. The statistically significant finding is that an increase in long-term exposure to $PM_{2.5}$ of just 1 μg/m^3 is associated with an 8% increase in Covid-19 mortality.

To put that mortality increase in perspective: the WHO guideline for annual average $PM_{2.5}$ is 10 μg/m^3, and the current US regulatory level is 12 μg/m^3. Further perspective may be gained from Table 1.2, where the annual values for the 10 most polluted cities in the world are over 100 μg/m^3. Clearly, the 8% mortality increase per unit increase in pollution is not linear beyond a certain range. In fact, one could even posit that the slope of the impact *must* be lower at the higher levels of $PM_{2.5}$.

Another clue to reconciling the mortality propensity in the most polluted cities is the range of PM levels in the Wu et al. study. The 3087 counties experience an average $PM_{2.5}$ of 8.4 μg/m^3 and are divided into two blocks: 1217 counties at <8.5 μg/m^3 and 1870 counties at >8.5 μg/m^3. On average, these are well below the regulatory level, although the second block almost certainly has some counties at levels exceeding the US regulatory threshold.

For a reasonability check, see the data plotted in Fig. 1.5, which show the drop in mean annual $PM_{2.5}$ levels since the start of the 21st century.[14] The decline has been steady, with the means falling below the national standard a decade ago and continuing to fall. Even the 85th percentile was below 12 μg/m^3 in 2018. Not surprisingly, this is consistent with the PM distributions in the counties studied in the Wu et al.'s paper. This also lends credence

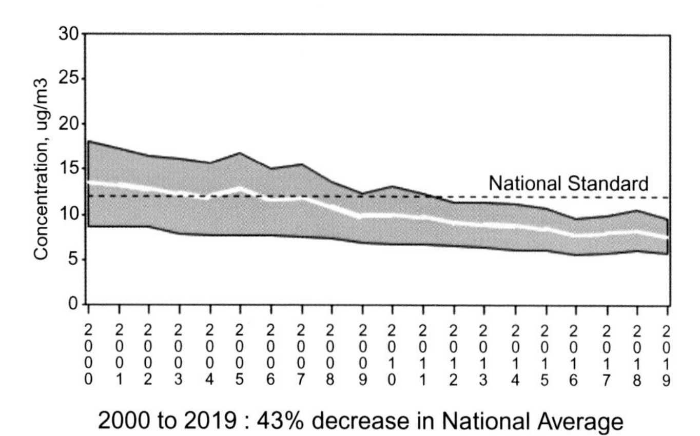

Fig. 1.5 $PM_{2.5}$ trend over the last two decades in the United States, with the mean shown in white and bounded by the 85th and 15th percentiles. *(Courtesy United States Environmental Protection Agency.)*

to the SARS study, where the API in every case was below the "safe" 100 threshold. But it does beg the question as to what one could expect in Covid-19-related mortality increase per unit increase in $PM_{2.5}$ in communities where the $PM_{2.5}$ level is an order of magnitude higher than the means in the Wu et al.'s study.

Another recent study estimated the mortality due to $PM_{2.5}$ over the period 1990–2010, without the Covid-19 complication. During these two decades, $PM_{2.5}$ dropped by 39%. This reduction in fine particulates avoided roughly 38,000 (38%) deaths when compared with the number expected if PM had remained at the 1990 level.[15] US mortality is only about 2% of the total for the world. This comports with the relatively low ambient pollution seen in Fig. 1.5. Also, their data show that a 7 $\mu g/m^3$ reduction avoided 38% of deaths. That in turn calculates to a 5.4% reduction in mortality for every 1 $\mu g/m^3$ reduction in $PM_{2.5}$. The same statistics, if calculated in the forward direction, predict an 8.6% increase in mortality per $\mu g/m^3$ increase in $PM_{2.5}$. For mortality associated with Covid-19, Wu et al. predict an increase of 8% with every 1 $\mu g/m^3$ increase in $PM_{2.5}$. These two separate studies over different but overlapping periods do appear to indicate that the Covid-19-related mortality attributed to $PM_{2.5}$ is about the same as an all-causes figure. However, other studies of mortality from PM alone show shallower slopes than in Zhang et al. One would expect a more severe outcome when the PM impact and that of the Covid-19 virus act in concert. One finding, even in the early days of the pandemic, has stood out; 94% of the severe outcomes occur when one or more of three comorbidities are present: hypertension, obesity, and diabetes. A working hypothesis would be that the elevated fine particle count increases the severity of the comorbidities and the virus simply adds to that. The virus may even be seen as yet another comorbidity tipping the scale toward mortality, rather than as a spectacular new mechanism.

At this writing, we are still very early in the understanding of Covid-19 and the factors affecting severity of outcomes. Because of the imperative to ameliorate or eliminate the effects of the pandemic, papers are being made public prior to peer review, and deductions drawn here, or in contemporary papers, are subject to reappraisal based on newer data. Furthermore, mortality trends such as in the Zhang paper are subjected to assumptions that result in uncertainties. But that is the state of the art today in predictions. Other models, also based on epidemiologic studies, predict as low as 1.5% change in mortality with every 1 $\mu g/m^3$ reduction in $PM_{2.5}$.[10] These issues, and how policy may be informed in the face of the uncertainties, are discussed in Chapter 14.

Transmigration of particulate matter

PM can be transported over long distances. This raises the specter of high PM counts even in regions with low local production. Most interventions are locally based, such as the effort that the city of Delhi undertook by legislating the conversion of public transport fuel from diesel to natural gas. Whereas this had a substantially positive impact, during the winter months, PM migrating from rice stalk burning in a neighboring state still leaves Delhi in the unenviable position of being in the top dozen most polluted cities in the world. Whereas the air quality in urban areas can be affected by other regions, the greatest impact is in rural areas, where local production is minimal.

The migration of PM across continents has been documented and is not in dispute. Aerosol transport from Asia to as far as North America has been reported.[16] In some areas, such migration constitutes a high fraction of the ambient aerosols. Long-range transport was determined in Taiwan in the winter monsoon period of 2000–01. The northeasterly winds of the winter monsoon in mainland Asia were identified as the pollutant carriers. The transported species were estimated to add 30 μg/m^3 to the ambient loading.[17] This is another piece of evidence demonstrating that levels double the WHO guidelines can be achieved from nonlocal sources.

A study in South America shows that on days correlated with biomass combustion in northern Brazil, the PM measured in the city of Londrina in southern Brazil was significantly increased.[18] Londrina, a relatively small city, was selected because ambient measurements have been made routinely for years and provide a good base case. The investigators measured black carbon with an aethalometer (see Chapter 6) at wavelengths 370 and 880 nm. The latter is the standard for estimating black carbon, and the shorter wavelength was believed to be strongly absorbing for the emissions from biomass combustion, whereas diesel combustion particles have weak absorption at these wavelengths. This focus on biomass combustion was driven by the fact that the expected source was the burning of sugarcane. Cane production increased in Brazil over the years of the investigation largely because of the national policy for displacing gasoline with ethanol. After the sugar is extracted for ethanol production, the bagasse (fibrous matter left behind) is combusted, often to produce local power. It is this combustion that many see as a downside to the otherwise positive move to use less petroleum-derived fuel.

Two days with different levels of biomass combustion were compared based upon satellite observations. The contribution of black carbon and $PM_{2.5}$ to the ambient concentrations varied from 70% to 87% depending on plume conditions. This underlines the importance of local pollution, with attendant health effects, occasioned by transmigration of species from distant sources of combustion. In this case, the source is combustion of biomass residue from a farming operation. This is the same class of sources as the rice stalk burning in India. A nonanthropogenic source would be wildfires. Some would argue, however, and we do so later in this chapter and again in Chapter 9, that wildfires too can have anthropogenic origins.

A recent study examined the effect on New York City from fires as far as 4000 km away.[19] The events occurred about 2 weeks apart in August 2018, each connected to a different source. One was a wildfire in western Canada and the other probably a controlled agricultural burn in the southeastern United States. Black carbon was measured by aethalometer and $PM_{2.5}$ by conventional filter capture. During the 2- and 3-day sample periods, the concentrations of black carbon and $PM_{2.5}$ were substantially greater than the baselines. $PM_{2.5}$ reached 35 μg/m^3 when compared with baselines of <10 μg/m^3 in the intervening periods. Black carbon showed even greater

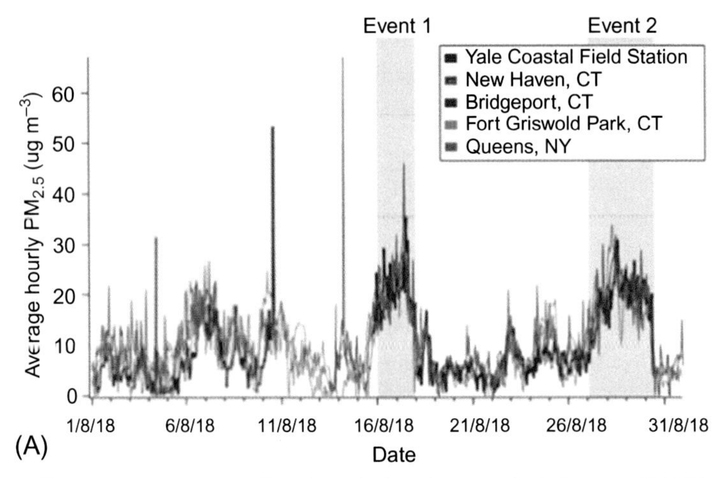

Fig. 1.6 $PM_{2.5}$ measurements at five sites during the month of August 2018. The *gray-shaded areas* represent the periods associated with the burns in western Canada (Event 1) and the southeastern United States (Event 2).[19] *(Courtesy Rogers HM, Gitto JC, Gentner DR. Evidence for impacts on surface-level air quality.* Atmos Chem Phys. *2020;20:671–682. https://doi.org/10.5194/acp-20-671-2020 (web archive link).)*

increases (Figs. 1.6 and 1.7). During those occasions, air quality advisories were issued for the metropolitan area.

Interestingly, the journal *Nature* picked up this paper and described it[20] under the headline "Passive Smoking: Canadian Fumes Engulf New York."

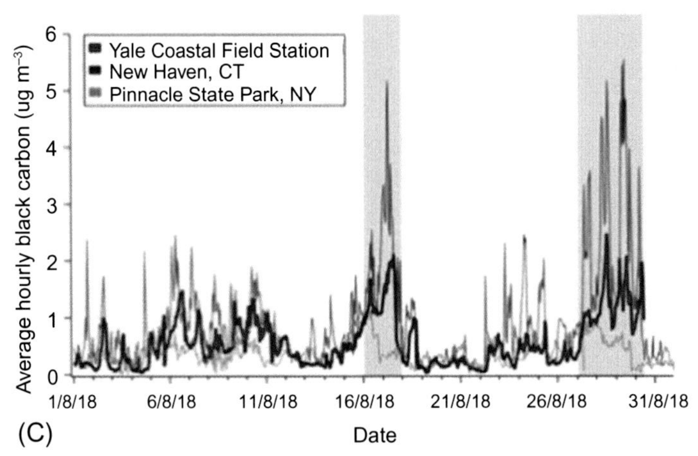

Fig. 1.7 Black carbon measurements at three sites on the same days as in Fig. 1.6, delineated by the *gray shading.*[19] *(Courtesy Rogers HM, Gitto JC, Gentner DR. Evidence for impacts on surface-level air quality.* Atmos Chem Phys. *2020;20:671–682. https://doi.org/10.5194/acp-20-671-2020 (web archive link).)*

Accompanying is an image of a forest fire (not in the original paper). Undoubtedly that makes better copy than the presumably deliberate agricultural burn. Even venerable journals fall prey to sensationalism, or more charitably, a nod to human interest.

Transmigration markers and climatic effects

Migration over distances is assessed through study of the chemistry of the species. Polycyclic aromatic hydrocarbons (PAHs) are produced when biomass containing organic molecules such as those in the lignin portion of trees is combusted. They are toxic and carcinogenic. One of the most toxic, benzo(*a*)pyrene (BaP), is commonly used as a marker for migration.

The concentration of BaP in locations distal from the source is greater than predicted by models of formation and destruction of the layers surrounding elemental carbon cores.[21] BaP sourced from woody biomass is adsorbed onto the elemental carbon emitted (soot) or other inorganic species comprising PM. Laboratory studies have shown that ozone in the atmosphere will oxidize the BaP in minutes to hours. This is inconsistent with the observation of long-lived BaP in transmigrating PM.

A recent study resolves this inconsistency.[22] The authors offer the hypothesis that BaP destruction by ozone is just one of two mechanisms determining the fate of BaP. The other is depicted in the lower panel of Fig. 1.8. The upper panel shows the conventional mechanism, where the BaP is not shielded from the ozone. The lower shows the shielded mechanism, where the ozone is impeded by a viscous layer of organic aerosol (OA), often a secondary organic aerosol (SOA), formed by organics from the surrounding ambience and designated as an SOA in the figure.

The two mechanisms have in common the *adsorption* of BaP, or any other PAH, onto the inorganic core, shown in this example as elemental carbon. It is also *absorbed* in the outer OA layer, labeled generically as OA in their paper and by us in this chapter. At first, this may be a primary organic aerosol, generated from the organics present in the combusted material. Subsequent reaction with organics in the ambience results in the formation of an SOA.

Reaction of ozone with the PAH is limited to the surface of the OA layer. In the unshielded model that has been assumed in the past to be the only mechanism, PAH molecules diffuse to the surface and react with the ozone. The shielded model is in play when the OA layer is viscous and reduces the mobility of the PAH. For a given OA composition, the

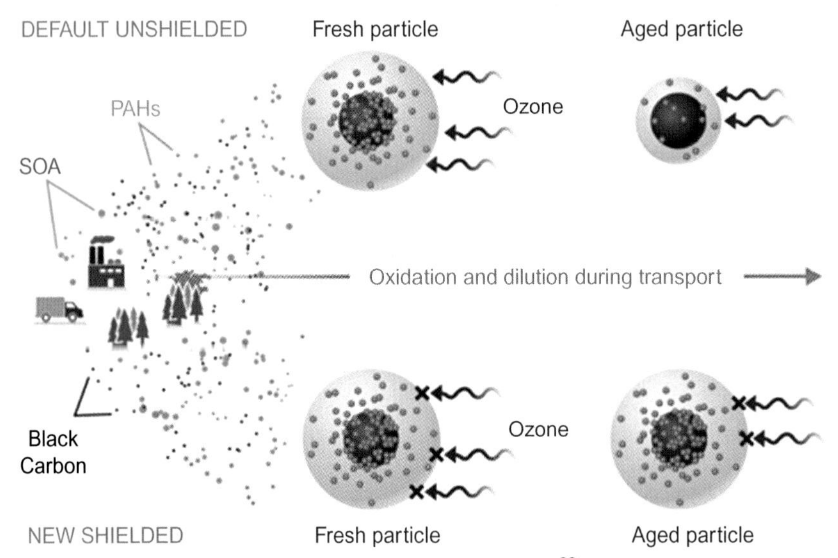

Fig. 1.8 The two mechanisms for BaP oxidation by ozone.[22] *(Courtesy PNAS, Shrivastava M, Lou S, Zelenyuk A. Global long-range transport and lung cancer risk from polycyclic aromatic hydrocarbons shielded by coatings of organic aerosol.* PNAS. *2017;114(6): 1246–1251.)*

viscosity is determined by the ambient temperature and humidity. Consequently, one would expect the shielded model to be more prevalent in the winter and spring and at higher altitudes.

Measured BaP at Asia sites is higher in winter and spring (Fig. 1.9). This is due in part to a near doubling of biofuel consumption in winter when compared with summer and in part to the cooler temperatures and lower humidity. The shielded model better matches the data than the unshielded model. The other measurements were at the Mt. Bachelor Observatory in the Cascade Range near the west coast of the United States. This location was selected because of the seasonal spring arrival of PM across the Pacific from Asia. Here, too, the shielded model performs better.

Lest this appear to be an exercise in fitting models to data, we will briefly discuss the health implications. Many of the PAH species, BaP in particular, are strongly implicated in cancer morbidity and mortality. Previous models using just the unshielded mechanism underpredicted mortality attributable to PAH loading in the ambience. Shrivastava et al. recalculated these, as shown in Fig. 1.10.

The unshielded model underpredicts the risk. The shielded model predicts ILCR values well above the threshold for acceptable risk. The worst

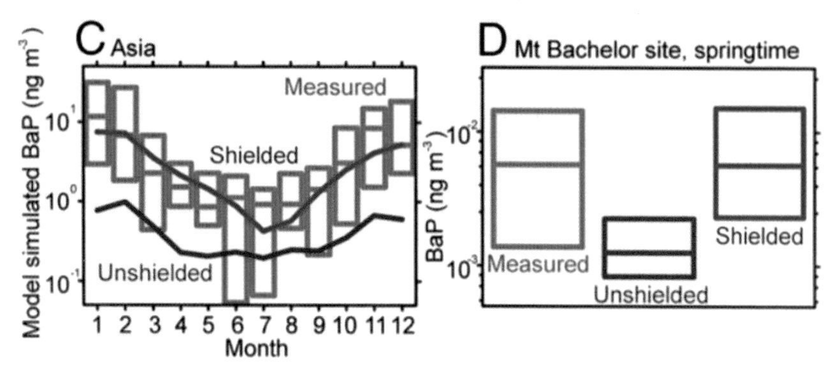

Fig. 1.9 Seasonal variation in BaP when compared with shielded and unshielded model predictions at 18 sites in Asia and the Mt. Bachelor site in the western United States. *Boxes* show the medians bounded by the 15th and 85th percentiles of site monthly means. *Green* is the measured points and *blue* and *red* are the medians of monthly means of the shielded and unshielded models, respectively.[22] *(Courtesy PNAS, Shrivastava M, Lou S, Zelenyuk A. Global long-range transport and lung cancer risk from polycyclic aromatic hydrocarbons shielded by coatings of organic aerosol.* PNAS. *2017;114(6): 1246–1251.)*

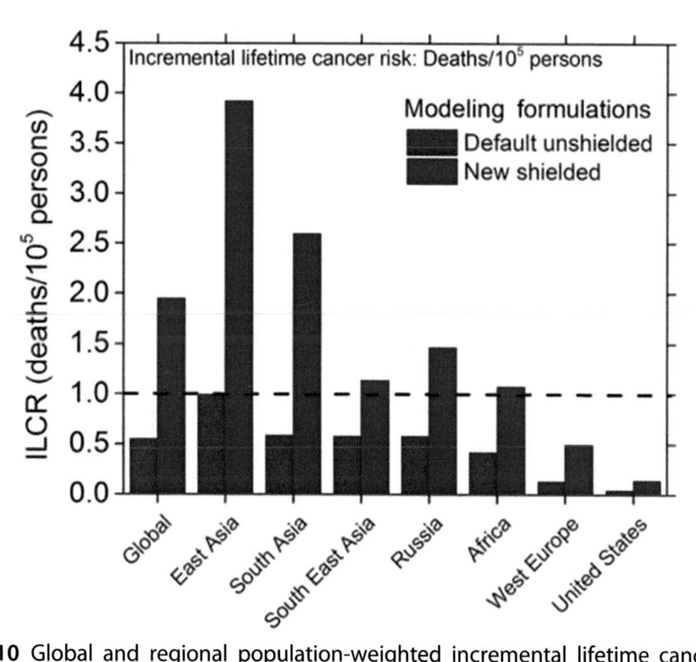

Fig. 1.10 Global and regional population-weighted incremental lifetime cancer risk (ILCR) attributable to polycyclic aromatic hydrocarbon mixtures using the two models. The *dashed line* represents the value over which the risk is considered significant.[22] *(Courtesy PNAS, Shrivastava M, Lou S, Zelenyuk A. Global long-range transport and lung cancer risk from polycyclic aromatic hydrocarbons shielded by coatings of organic aerosol.* PNAS. *2017;114(6):1246–1251.)*

areas are East Asia and South Asia, with most of the latter being in India. The older model would have given a false sense of security in this case. Europe and the United States have low ILCR predictions even with the new model, but lower margins for error than with the default model.

Close-up: Shedding light on the Covid-19 virus

If enough of the light is at ultraviolet wavelengths, the virus will die. This light, however, is an attempt to explain some of the science behind the virus and its effects. I (V.R.) fully expect the reader to obtain fact or inference checks from physician scientists but think this will endure scrutiny.

A general caution is that very few sites ought to be relied upon without verification. Reputable sites include the National Institutes of Health and in particular the National Institute of Allergy and Infectious Diseases, the Centers for Disease Control, and the WHO. Other sources are sites at top medical schools such as at Johns Hopkins, Stanford, and Harvard.

First the nomenclature. Covid-19 is the *disease* resulting from the virus. The *virus* is from the general family of coronaviruses, with this variant being named SARS-CoV-2. SARS stands for severe acute respiratory syndrome. The name being a bit of a mouthful; the WHO often refers to it as the Covid-19 virus. It is related to those responsible for the outbreaks of SARS in 2002–04 and Middle East Respiratory Syndrome in 2012.

In common with other coronaviruses, these are spherical, with protein spikes sticking out about 12 nm. Resemblance to a crown informs the corona name. They also have a striking resemblance to that fearsome medieval weapon the mace. In size they are reported to be in the range of 50–150 nm, which places them roughly in the ultrafine classification of airborne aerosols. However, deposition fractions in various parts of the respiratory tract cannot be presumed to be similar to those of PM, even those coated with organic molecules.

Fig. 1.11 is a virus isolated from a Covid-19 patient in the United States. The spiky proteins attach to receptors in human cells. The mechanism is not unlike a lock and key. The key of the virus protein needs a receptor lock to attach to so as to enter the cell. Another analogy is docking of a spaceship to a space station. Once docking happens, the virus can enter the cell. Then, it can replicate, and the disease will be well on its way. Recent research has shown that the receptor for SARS-CoV-2, which causes Covid-19, is the same as that for the SARS virus. That is the good news because we know a lot about the original SARS. The not so good news is that the binding affinity (the ease with which it binds) for this virus is 10–20 times greater than for the original SARS.[23] This could explain why the human-to-human spread appears to be greater than was noted in the SARS outbreak. Furthermore, despite the similarities in the structure and sequence of the protein

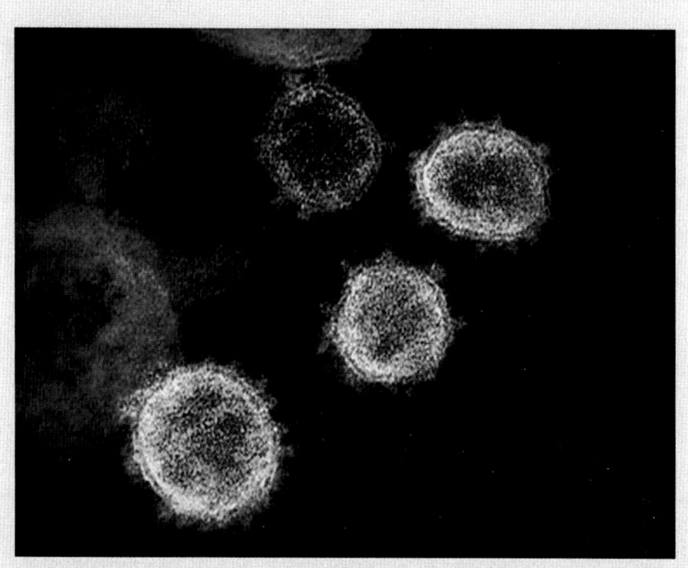

Fig 1.11 Transmission electron microscopy image of a SARS-CoV-2 virus. *(Courtesy NIAID-RML source NIH Research Matters.* Novel coronavirus structure reveals targets for vaccines and treatments; *2020. https://www.nih.gov/news-events/nih-research-matters/novel-coronavirus-structure-reveals-targets-vaccines-treatments. Retrieved 16 May 2020.)*

spikes of the two viruses, three antibodies developed for SARS did not bind effectively to the SARS-CoV-2 protein spike.[24]

A feature of SARS-CoV-2 is that it is enveloped by a lipid (fat) layer (the "crown" protein spikes extend beyond the lipid layer). In this aspect, the structure is like that of influenza viruses and other coronaviruses (and unlike the diarrhea-inducing rotavirus). This is fortunate because soap and water will kill it. Soap has a hydrophilic head and a lipophilic tail. The tail penetrates the lipid layer and pries it apart, thus leading to the destruction of the viral genes, with all the fragments being washed away by the water. This mechanism of action underlies the most important public health guideline for minimizing spread, washing of hands in soap and water for at least 20 s, taking care to wash between the fingers. Hand sanitizers are believed to be effective if they contain at least 60% ethanol. They too remove the lipid layer and cause the disintegration of the virus, but the debris does not get washed away.

Except for the hand-washing discussion, I will not get into disease avoidance. For the rest, you need to go to one of the reputable sites. But I will note that my limited examination of the literature shows a flurry of scientific activity on several fronts. These include studies of the immune response, development of a test to verify the presence of antibodies in the United Kingdom, testing of intensity reduction drugs (e.g., Tamiflu for influenza), and research on the ultimate prize, a vaccine. Keep in mind that folks are rushing to publish, and so findings may be subjected to revision. The immune response study reference above is based on a single patient but still instructive. With all the stuff out there, *caveat emptor*!

References

1. Brimblecombe P. The Big Smoke: A History of Air Pollution in London since Medieval Times. London: Methuen; 1987.
2. Cao J, Chow JC, Lee FSC, Watson JG. Evolution of PM2.5 measurements and standards in the U.S. and future perspectives for China. *Aerosol Air Qual Res*. 2013;13:1197–1211. https://doi.org/10.4209/aaqr.2012.11.0302.
3. Miller FJ, Gardner DE, Graham JA, Lee RE, Wilson WE, Bachmann JD. Size considerations for establishing a standard for inhalable particulates. *J Air Pollut Contr Assoc*. 1979;29:610–615.
4. Lippmann M. Human respiratory deposition of particles during oronasal breathing. Atmos Environ. 1984;18:1038–1039.
5. Chow JC, Watson JG, Mauderly JL, et al. Critical review discussion—health effects of fine particulate air pollution: lines that connect. *J Air Waste Manage Assoc*. 2006;56:1368–1380.
6. Pope III CA, Dockery DW. Critical review: health effects of fine particulate air pollution: lines that connect. *J Air Waste Manag Assoc*. 2006;56:709–742.
7. United States Environmental Protection Agency. National ambient air quality standards for particulate matter: final rule. *Fed Regist*. 1997;62:38651–38760. http://www.epa.gov/ttn/amtic/files/cfr/recent/pmnaaqs.pdf.
8. World Health Organization. *Air pollution levels rising in many of the world's poorest cities*; 2016. Available from http://www.who.int/news-room/detail/12-05-2016-air-pollution-levels-rising-in-many-of-the-world-s-poorest-cities. Retrieved 2 March 2020.
9. Vincent JH, ed. *Particle Size-Selective Sampling for Particulate Air Contaminants*. Cincinnati: ACGIH Signature Publications; 1999.
10. Hayes RB, Lim C, Zhang Y, et al. PM2.5 air pollution and cause-specific cardiovascular disease mortality. *Int J Epidemiol*. 2019;dyz114. https://doi.org/10.1093/ije/dyz114.
11. Cui Y, Zhang ZF, Froines J, et al. Air pollution and case fatality of SARS in the People's Republic of China: an ecologic study. *Environ Health*. 2003;2:15.
12. Friedman L. New research links air pollution to higher coronavirus death rates. *New York Times*. April 17, 2020. https://www.nytimes.com/2020/04/07/climate/air-pollution-coronavirus-covid.html. Retrieved 15 May 2020.
13. Wu X, Nethery RC, Sabath BM, Braun D, Dominici F. Exposure to air pollution and COVID-19 mortality in the United States. *medRxiv*. 2020. https://doi.org/10.1101/2020.04.05.20054502.
14. United States Environmental Protection Agency. *Particulate matter (PM2.5) trends*; 2020. https://www.epa.gov/air-trends/particulate-matter-pm25 trends. Retrieved 16 May 2020.

15. Zhang Y, West JJ, Mathur R, et al. Long-term trends in the ambient PM2. 5-and O3-related mortality burdens in the United States under emission reductions from 1990 to 2010. *Atmos Chem Phys.* 2018;18(20):15003.
16. Lin M, Fiore AM, Horowitz LW, et al. Transport of Asian ozone pollution into surface air over the western United States in spring. J Geophys Res. 2012;117:D00V07.
17. Lin C-Y, Liu SC, Chou CC-K, et al. Long-range transport of aerosols and their impact on the air quality of Taiwan. *Atmos Environ.* 2005;39(33):6066–6076.
18. Martins LD, Hallak R, Alves RC, et al. Long-range transport of aerosols from biomass burning over southeastern South America and their implications on air quality. *Aerosol Air Qual Res.* 2018;18:1734–1745.
19. Rogers HM, Gitto JC, Gentner DR. Evidence for impacts on surface-level air quality. *Atmos Chem Phys.* 2020;20:671–682. https://doi.org/10.5194/acp-20-671-2020.
20. Passive smoking: Canadian fumes engulf New York. *Nature.* 2020;578:11 (6 February).
21. Sehili AM, Lammel G. Global fate and distribution of polycyclic aromatic hydrocarbons emitted from Europe and Russia. *Atmos Environ.* 2007;41(37):8301–8315.
22. Shrivastava M, Lou S, Zelenyuk A. Global long-range transport and lung cancer risk from polycyclic aromatic hydrocarbons shielded by coatings of organic aerosol. *PNAS.* 2017;114(6):1246–1251.
23. Wrapp D, Wang N, Corbett KS, et al. Cryo-EM structure of the 2019-nCoV spike in the prefusion conformation. *Science.* 2020. https://doi.org/10.1126/science.abb2507. pii: eabb2507. [Epub ahead of print] 32075877.
24. NIH Research Matters. *Novel coronavirus structure reveals targets for vaccines and treatments*; 2020. https://www.nih.gov/news-events/nih-research-matters/novel-coronavirus-structure-reveals-targets-vaccines-treatments. Retrieved 16 May 2020.

CHAPTER TWO

Principal sources of PM

The primary focus of this book is the impact of anthropogenic particulate matter (PM) on the health of people. Impact on the climate is clearly an important consequence of airborne particulates and that impact in turn certainly affects people. This chapter discusses the combustion sources that are acknowledged as the worst actors with respect to health impact.[1] This emphasis notwithstanding, we do discuss natural sources when they act in concert with the anthropogenic ones, such as in Chapter 5, where naturally sourced dust is a contributory player. Forest fires are mentioned, in part because their frequency is on the rise; that too is a combustion source, and a high proportion are thought to be caused by human intervention.[2] Some people believe that the increased prevalence is a consequence of global warming, therefore, that natural PM is likely to increase even as anthropogenic sources are being curtailed by policy and engineered solutions.[3] The close-up at the end of the chapter is a nod toward another natural source, desert dust.

Internal combustion engines

Internal combustion engines are devices in which combustion products expand to propel an object to perform work. In an automobile engine, the object is a piston, and the work is mechanical movement of parts and ultimately the wheels. Curiously, a rifle may also be viewed as an internal combustion engine, with the object propelled being a bullet. The principal fuels are gasoline and diesel, combusted using the oxygen in air. In both cases, the fuel-air mixture is ignited when a designed level of compression is achieved. In a gasoline engine, ignition is by a high energy spark, and in a diesel engine, it is by the heat generated by compression—thus the descriptors spark ignition and compression ignition. As we will see in Chapter 11, an engineered solution to reduce diesel-based PM is substitution with natural gas. In that case, the fuel must be spark ignited. When the substitution is with clean-burning dimethyl ether, normal compression ignition is sufficient.

Particulates Matter
https://doi.org/10.1016/B978-0-12-816904-9.00007-6

In both types of engines, the combustion conditions dictate the nature and quantity of the emissions. The higher the compression prior to ignition, the greater the work produced. High compression engines are more efficient, providing more distance driven per unit of fuel. Accordingly, the emissions per mile are less. Diesel fuel has about 10% more energy density than gasoline, but diesel engines are more efficient by about 30% because of a combination of greater energy density and higher compression. Typical diesel compression ratios are about double those of standard gasoline engines. Increased efficiency notwithstanding, diesel engines are far more polluting than gasoline engines if the mass of PM per unit volume of emissions is the criterion.

Lower combustion temperatures yield lower oxides of nitrogen (NO_x) emissions because the oxidation of the nitrogen in the air mixture is accelerated at higher temperatures. Some of the amelioration schemes described in later chapters take advantage of this feature by providing means that lower the combustion temperature. In fact, since the introduction of fuel injection in gasoline engines (as opposed to fuel and air mixing in a carburetor), there has been increasing reliance upon the cooling effect of fuel vaporization. The best results are achieved with direct injection, where the fuel goes directly into the cylinder, rather than into a preliminary port (ported injection). This has two advantages: atomization of the fuel, making it easier to ignite, and absorption of heat from the cylinder to cause the liquid to vaporize. The higher the latent heat of vaporization of the fuel, the greater the effect. The 10% or so ethanol currently blended in gasoline gives an additional boost because the latent heat of ethyl alcohol is 2378 British Thermal Units per gallon (BTU/gallon), when compared with 900 BTU/gallon for gasoline. In a gasoline engine running with a stoichiometric proportion of oxygen (just enough to combust the fuel molecules), fuel injection will reduce the temperature in the cylinder by about 22°C. The cooling effect allows more compression while still not overheating to cause premature ignition (knocking). Consequently, direct injection gasoline engines run at higher compression ratios. This is likely the reason that the newer Mazda's SkyActiv engines operate at a ratio of 12 on regular octane gasoline, when compared with a normal 8.5. The gas mileage is correspondingly higher. If methanol is part of the 10% alcohol, the cooling effect is even greater because the latent heat of evaporation of methanol is 3340 BTU/gallon. The United States does not permit this (the Open Fuels Act, legislation to allow it, has been on idle in Congress since 2009), but Australia does with its GEM fuels (gasoline, bio-ethanol, and bio-methanol).

Diesel engines

We concentrate on diesel engines in part because the focus is on PM and associated emissions such as NO_x and organic molecules. Gasoline engines emit a fraction of the PM of diesel, although there is some evidence that the disparity is not as great when ultrafine particulate count (as opposed to mass) is the parameter.[4] In fact, one study showed more emissions from gasoline engines when compared with diesels with particulate filters.[5] The authors caution that the bulk of diesel vehicles in Europe in 2016 did not have such filters. The same could be said for Asia. But that result does point to what is possible with ultra-low sulfur diesel as the fuel and particulate filters in the exhaust system. With all this evidence at hand, we feel that the emphasis here on diesel combustion is valid, at least until the legacy fleet becomes a small fraction of vehicles on the road.

Gasoline-consuming two and three wheelers are demonstrably heavy polluters even by conventional measures,[6] and they are discussed separately. Two wheelers are especially prevalent in developing economies. They are affordable precursors to car ownership. A common sight in India is a motorcycle with a small entire family on it. There is no luggage, though. These economies are also where other sources such as wood burning are major contributors to atmospheric aerosols.

Chemistry of diesel

Diesel is distinguished from the other two principal fuels derived from crude oil by the carbon-to-hydrogen ratio. Crude oil is a mixture of molecules and can be considered to have the formula C_nH_{2n+2}, where n is an integer. The higher the proportion of molecules with larger values of n, the heavier the crude. Practically, the American Petroleum Institute (API) defines weight by the parameter API gravity. It is an inverse measure of the specific gravity with respect to that of water. When the API gravity is less than 10, the oil is heavier than water and will sink in the mixture. Very few crudes have this character, and they are classified as extra heavy. In Canada they are called bitumen. The lightest crude currently available in quantity is shale oil, with API gravity between 40 and 50. The simplest form of refining crude is fractional distillation. The fractions at different boiling points are collected. Of the three transport fuels, the earliest condensing is gasoline, followed by jet fuel, then by diesel. Each of these is a mixture with the approximate formula

$$C_nH_{2n+2} \tag{2.1}$$

with n spanning a range and most of the molecules roughly in the ranges.

$$\text{Gasoline } 5-9, \text{ Jet fuel } 9-14, \text{ and Diesel } 12-19. \tag{2.2}$$

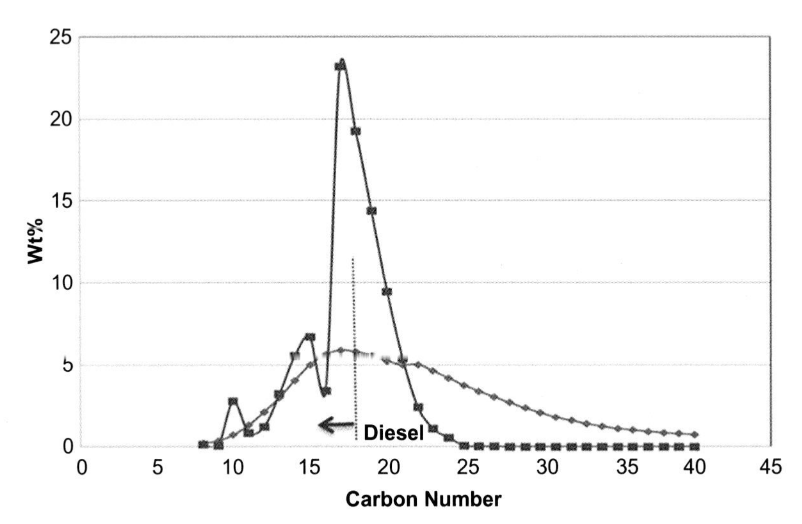

Fig. 2.1 Percentage of distribution of molecules in diesel fuel. *(Courtesy: RTI International, private communication of internal report.)*

To complicate matters further, the actual distribution of molecules is dependent upon the method of manufacture. The most startling difference is when diesel is synthesized from natural gas. This is commonly done by Fischer-Tropsch (FT) synthesis, in which syngas ($CO + H_2$) is polymerized to yield a mixture of molecules approximating diesel. The largest molecules comprise wax and are separated, leaving a fuel with superior properties in parameters such as zero sulfur content and high cetane number. Different FT processes yield somewhat different distributions of molecules. The reaction being highly exothermic, the distinguishing feature is the means by which the exotherm is controlled. The Greyrock process claims to produce no wax,[7] accomplishing this with a unique catalyst.

Fig. 2.1 plots the parameter n (carbon number) against the weight percent distribution for a standard diesel in blue and an FT diesel in red. This FT process controls the exotherm sufficiently to produce almost no molecules with a carbon number greater than 25. Such a product is preferred because the larger molecules have a greater propensity for soot production in the exhaust. The Greyrock process also produces a liquid with a distribution similar to the red line.

The principal emissions of interest are compounds of sulfur, organics, PM, and NO_x. The sulfur is entirely predicated on the original fuel having the element in some form, and this is increasingly being regulated downward. The Euro V standard requires the sulfur to be below 10 ppm. This is a substantial decrease from the Euro IV standard of less than 50 ppm. The Euro standards are seen worldwide as the ones to emulate. But the cost of achieving the higher standards, especially from sour heavy oil sources, has slowed down the adoption in many countries. Diesel from natural gas conversion (which we refer to as FT diesel, even though the process used may not necessarily be FT) will have zero sulfur. All the others are much more dependent on operating conditions, although FT diesel will have zero aromatics, and so the organics emitted will be more benign.

Unlike in spark ignition engines, the combustion in diesel engines is substantially nonhomogenous. The fuel is injected directly into the cylinder when the desired compression ratio is achieved. This is also the time at which the compressed air is at a high temperature, due in large measure to the frictional heat of compression. The fuel ignites but is not necessarily well mixed with the air prior to ignition. This inhomogeneity allows pockets of incomplete combustion, even though the efficiency of combustion in modern diesel engines is at least 98%.[8] The cause of incomplete combustion is local oxygen deprivation (below stoichiometric levels) or relatively low temperatures. The product in these cases can include carbon monoxide (CO) and uncombusted hydrocarbons, often aromatics.

Oxides of nitrogen are produced at relatively high temperatures and fuel mixtures with oxygen greater than stoichiometric or lean. Engine people use the term equivalence ratio that is 1.0 for stoichiometric and less than 1.0 for a lean mix. The compounds generally comprise nitrogen monoxide (NO) and nitrogen dioxide, the latter being more toxic. But both play a significant role in ozone formation, in the presence of CO or hydrocarbons and with a photochemical step, as discussed in Chapter 3. Ozone is implicated in making the organic coating on secondary organic aerosols even more toxic, in part through the production of reactive oxygen species, as discussed in Chapter 4.

Two-stroke engines

The two-stroke engine (Fig. 2.2) is the workhorse of two-wheel vehicles such as motorcycles and autorickshaws (two-passenger taxis in Asia that are three wheelers). As a class, gasoline-powered two-stroke engines, whether in two wheelers or three wheelers, are decidedly worse on emissions than other vehicles.[6] In recognition of this, many countries are attempting to retire them in favor of more costly four-stroke engines, with varying degrees of success. As legacy vehicles will continue to be an issue, two-stroke engines are addressed in this chapter.

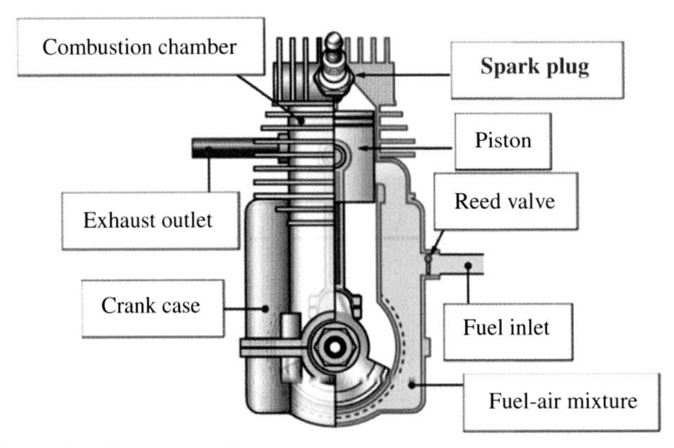

Fig. 2.2 Schematic of a two-stroke engine.

The engine has no valves, with the piston acting to open and shut the intake and exhaust ports that are drilled into the cylinder walls. In Fig. 2.2, the piston is shown shutting both the exhaust and intake ports. When the piston moves up high in the compression stroke, covering the exhaust port, the fuel-air mixture is compressed and ignited by the spark plug. The compression stroke also creates a vacuum in the crankcase, shown in yellow. This vacuum opens the reed valve and sucks the fuel-air mixture in from the carburetor. Ignition pushes the piston down, and the connecting rod causes a rotation in the lower assembly that drives the wheels. As the piston moves down upon firing, the inlet port, on the right in the figure, opens. The fuel-air mixture is sucked into the cylinder by the vacuum created and drives out the exhaust gases to the left. This new fuel-air mixture now awaits the compression stroke and combustion. But in pushing out the exhaust gases, inevitably some new fuel escapes into the exhaust. The exhaust of unburned hydrocarbons is one reason for these engines being worse on emissions. The other, more obvious reason is that lubricating oil is mixed with the gasoline. The purpose is to lubricate the crankshaft, connecting rod, and cylinder walls. Upon combustion, the PM count is much higher than it would be from just gasoline. In many such vehicles, visible black smoke in the exhaust is common.

Coal combustion

Trace metallic elements are present in coal. The concentrations range from low single digit parts per million up to a few hundred. Zinc tends to be in the highest concentration, followed by lead, copper, and nickel. Even very low single-digit concentrations are a source of nucleating agents. The vapor phase of the compound in air is known as the solute. Homogeneous nucleation occurs when the solute concentration exceeds the saturation, in which case it is described as supersaturated. The saturation of any solute A is described by the equation

$$S = p_A / p_{AS}, \tag{2.3}$$

where p_A is the partial pressure of A and p_{AS} is the saturation vapor pressure of A. In the supersaturated state, the value of S will exceed 1, and this condition is required for homogeneous nucleation to occur. However, for the particle to remain stable, the size has to exceed the "critical size." Once it does so, it immediately begins to grow through coagulation and reaction with the coproduced organics, as described in Chapter 4.

Uranium is also present and ranges from about 8 ppm to 20 ppm in the U.S. fly ash (almost a tenth of that in the coal). The United States Geological Survey (USGS) conducted a study of uranium in coal and fly ash. They concluded two points of interest to us: uranium preferentially concentrates in the fine portion of fly ash, and it is found in both organic and mineral (oxide) form.[9] Thorium, by comparison, is only in mineral form. The importance of this is more for fly ash disposal than for airborne particulates. Fly ash is often disposed of in ponds that famously overflowed in a Tennessee flood caused by a dam breach a few years ago. More prudently, it is used as a component of concrete, in part because the major constituents are mixed oxides of silicon and aluminum that are very cementitious. A concern has always been the leachability of heavy and radioactive metal compounds. The USGS investigation concluded that much of the uranium was in the glassy state, as shown in Fig. 2.3, reproduced here from Fig. 3 in their report. The fission track radiograph indicates that the uranium is in the glassy phase, which surrounds a hollow middle, typical of cenospheres in fly ash. The presence of the glassy phase likely means that it is present as an oxide that is a component of a mixed oxide compound comprising oxides of silicon and aluminum with a small proportion of oxides of uranium. Such mixed oxides, especially in a glassy crystalline state, can be expected to be stable with respect to leaching by ground water.

The other finding, that uranium may preferentially be in the fine particles, is more significant. The fine particle is not defined by the USGS publication, but the nominal definition is PM 2.5, which would include the ultrafine component. Both fine and ultrafine particles are contributors to

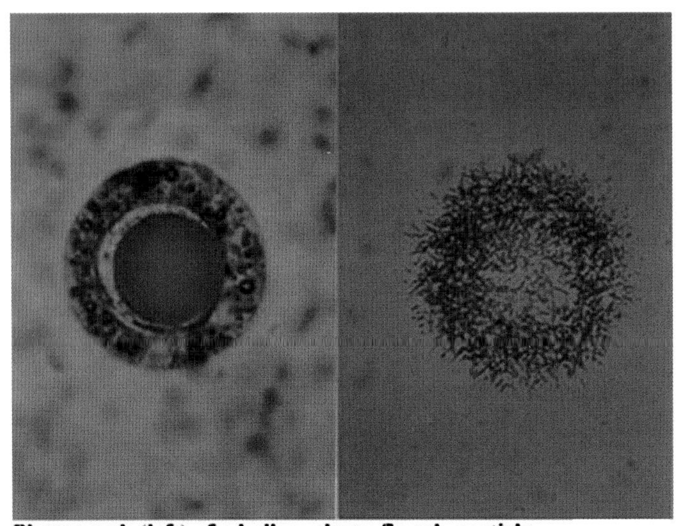

Fig. 2.3 Uranium in fly ash is present in the glassy phase. *(Courtesy: United States Geological Survey.)*

mortality, as opposed to the coarser PM10 that is more connected to morbidity. The ultrafine particles, with their ability to enter cells, is even more toxic if the payload includes radioactive elements. In fact, this concept is used in medical research, where nanoparticles carry therapeutic or diagnostic payloads.

Several other investigators have reported partitioning of metals into the ultrafine component of fly ash. Partitioning is a term that quantifies the relative distribution of a species in different phases, often in equilibrium. High specific gravity oil, especially that originating in Venezuela, contains relatively high concentrations of the heavy metal vanadium. One refining technique, the ROSE (residuum oil supercritical extraction) process, is to mix the crude with a solvent comprising light molecules such as propane and butane. The vanadium preferentially partitions into the asphaltene precipitate, and the remaining oil (and solvent) is virtually free of the impurity. The solvent is recovered for reuse.

Here, partitioning merely refers to the relative abundance in particles of that size. The partitioning is mediated by operating parameters such as combustion temperature and oxygen concentration. Sui et al. reported that at increasing temperatures and oxygen concentrations, the concentrations of Si, Fe, Cu, Zn, and Pb are 2–3.5 times greater in the size fractions below 0.4 μm.[10]

If indeed metal compounds are enriched in the ultrafine particles, the relative toxicity of the species matters. We have already discussed the case of uranium. More commonplace toxic elements are Hg, As, Se, Cd, and Pb, listed in ascending boiling points of the elements in metallic form. The presence of these in ultrafine PM would be disproportionally hazardous. Because the initial nuclei formed are less than 20 nm in size, even small concentrations could produce numerous nuclei. The partitioning of As, Se, and Cd in eastern U.S. coals into the submicron particle range was reported by Seames and Wendt.[11] They also noted that, in some coals, the presence of Ca caused some of these elements to form a complex with the Ca, resulting in them being found in the larger particle sizes. This underlines the fact that all coals are different, not just in concentrations of the elements but also in the mechanisms that determine the fate of constituents. But all the evidence supports the belief expressed in this book, that metal droplets are key nucleating agents for PM and not just for coal combustion.

Biomass combustion for cooking and heating

The term biomass covers a lot of ground, ranging from wood for cooking to industrial operations such as charcoal production to trees in forest fires. By far, the greatest volume in PM production from biomass comes from wood-burning cookstoves. This form of emission is the most difficult to define, in part because of the variability of sources. Even within a given source, combustion parameters determine the nature of the particles. Whereas this is certainly the same, in principle, for the other two major sources, internal combustion engines and coal, they are used under industrial conditions with tight controls. Biomass burning is practiced with loose controls and a much greater span of types. The nature of biomass is seasonal. In India, the three main sources for cooking and heating are firewood, cow dung, and crop residues, in that order, but agroclimatic influences cause the proportions to vary.[12] Source apportionment requires regional data, acquired primarily for input to climate models. Understanding the nature of PM for the impact on health also requires these data.

Firewood has many origins but may broadly be classified as hardwood and softwood. Nomenclature notwithstanding, the physical property of hardness is not always related. Balsawood, famously light, is classified as a hardwood. For our purposes, the key differences lie in the composition of the organic molecules. As discussed in Chapter 4, polycyclic aromatic hydrocarbons (PAHs) are important nucleating agents, together with black carbon (soot). The propensity for forming PAHs is linked to the existence of double- and triple-bonded carbon in hydrocarbons in the fuel. Fuel composition is, therefore, a key determinant of particle size and character. Another important variable is the burning condition, ranging from smoldering to fiercely flaming. Smoldering occurs when most of the volatile constituents have been released (not unlike in charcoal production). In these conditions, temperature remains low and uncombusted hydrocarbons will be present. These condense onto prevalent nuclei, with PAHs not among them because they need higher temperatures for formation. Intense high temperature fires tend to be oxygen rich and result in increased particle count. This is illustrated in Fig. 2.4 that describes the particle count and size distribution for wood burned with different oxygen availability.

The particle count is over double at 11.6% O_2 when compared with the low oxygen condition. In the oxygen starved condition the mode shifts to

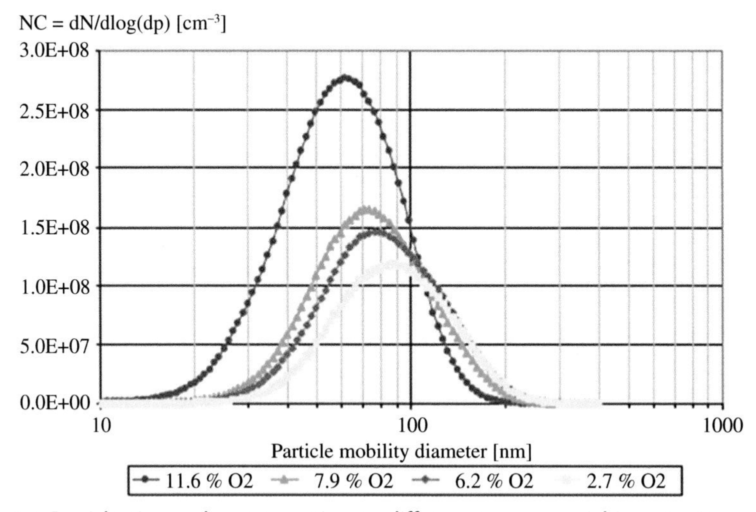

Fig. 2.4 Particle size and concentration at different oxygen stoichiometry in a wood chip boiler. *(Courtesy: Gaegauf C, Wieser U, Macquat Y. Field investigations of nanoparticle emissions from various biomass combustion systems. In 5th Conference on Nanoparticle-Measurement, Zürich, 2001.)*

larger sizes as well, likely due to unburnt organics depositing on the nuclei and growing them. One would expect this general behavior for most combustion processes, not just for biomass.

Dung as a fuel (dried and compacted fecal waste from animals, such as cows, buffalos, and yaks, often bound together with a fibrous material of convenience) is prevalent in developing nations. Whereas the woods have their variations, dung is a different material altogether and has distinctly different combustion characteristics. It tends to be on the smolder side of burn. This is evidenced in the particle size, chemistry, and count (Fig. 2.5).

The same study that produced the data in Fig. 2.5 also tested combustion with two types of firewood, mango and deodar. The results show a completely different distribution, with smaller sizes, although also with three modes (Fig. 2.6).

Crop residue burning

This topic deserves special mention even though, as noted above, it comprises a significant biomass source for cooking and heating. The crop residue for cooking is from stalks and stems of agricultural products such as rice and wheat that are produced in developing nations by scything and subsequent removal of the kernels. An especially important residue in

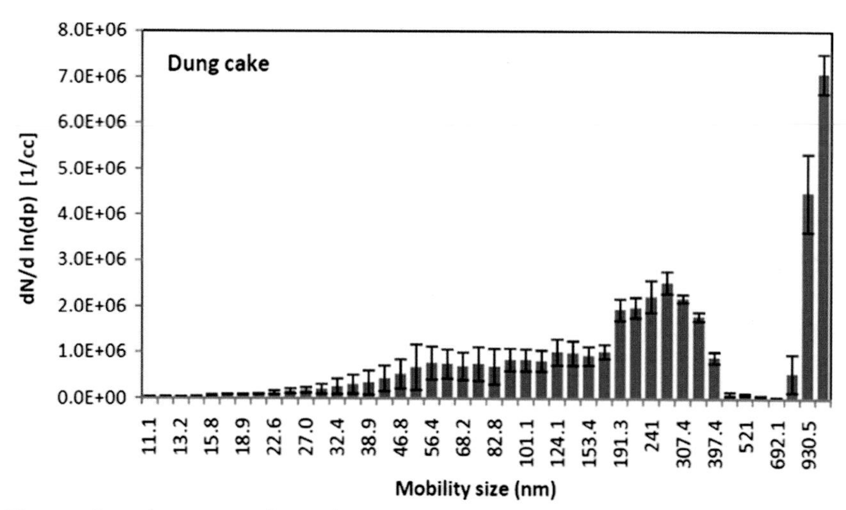

Fig. 2.5 Particle count and size distribution in combustion of dung in a cook stove. *(Courtesy: Tiwari M, Sahu SK, Bhangare RC, Yousaf A, Pandit GG. Particle size distributions of ultrafine combustion aerosols generated from household fuels. Atmos Pollut Res 2014;5 (1):145–150. doi:10.5094/APR.2014.018.)*

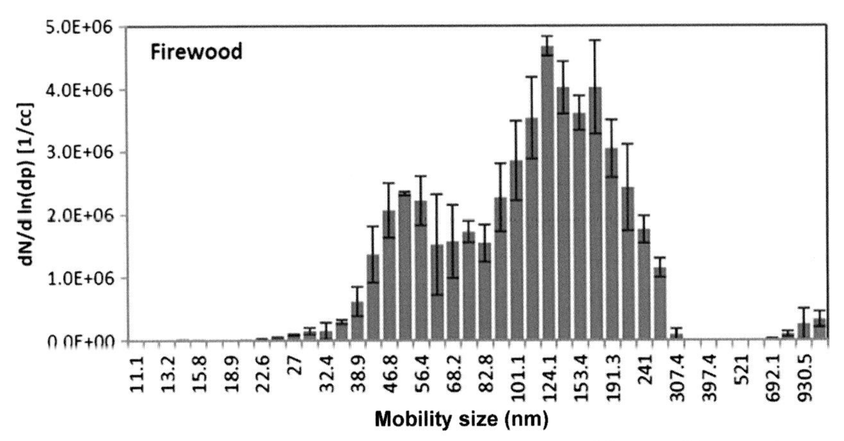

Fig. 2.6 Particle count and size distribution in combustion of firewood in a cookstove. *(Courtesy: Tiwari M, Sahu SK, Bhangare RC, Yousaf A, Pandit GG. Particle size distributions of ultrafine combustion aerosols generated from household fuels. Atmos Pollut Res 2014; 5(1):145–150. doi:10.5094/APR.2014.018.)*

some countries, notably India and Brazil, is bagasse, the fibrous portion of sugarcane left behind when the juice is pressed out. It is often combusted for power but must be dried first.

The source discussed here is a residue that is left in the ground after machine topping of the food-related portion such as wheat kernels. In southern India, the stalks are hand scythed and threshed for the rice, leaving the rice straw for biomass cooking or heating. In the northern Indian states of Punjab and Haryana, the wealthier farmers can afford machine cutting. The current practice removes just the tops, and the stalks left behind are burnt in the field. This is less costly than scything for sale or use. The PM produced in this combustion is credited with being the single greatest contributor to the poor air quality in the capital city of Delhi, downwind from the fields. Delhi has the dubious distinction of being in the top five of the world's cities in PM 2.5 count. Although the causality with rice stalk burning is disputed by some, it certainly is a principal source.

Dust storms

Dust storms, also known as sandstorms, are natural events. They may be mediated by human behavior in a process known as desertification. They mostly originate in desert areas. They often travel swiftly, at speeds in excess of 60 mph. The only reason for mention in this book is that, despite being of natural origin, they do contribute to airborne particulates. In the subSaharan region, the World Health Organization has cited dust storms as the cause of outbreaks of meningococcal meningitis. The particle sizes are mostly in the PM 10 region, with some observations of modes of 2.0 μm for particles staying airborne and 3.5 μm for those tending to settle. Given the regional nature of the effects, we leave the subject with only this brief mention.

Close-up: The time a dust storm wasn't one

I was born in Jodhpur (India); at the time a sleepy backwater with pretensions based largely on being the seat of a monarchy with wealth and influence. When the British departed in 1947, the maharajah at the time very nearly elected not to join the newly independent Republic of India. Not sure what his alternatives were. As a Hindu king, joining the new Muslim country of Pakistan a hundred or so miles to the west would not have made sense. Not sure what he was thinking. Then again, the thought processes of monarchs and presidents can often be puzzling. In any case, he remained with the rest. This was just as well because I was 3 years old at the time and not equipped to deal with the fallout of changing regimes. Not sure one ever is.

Jodhpur is also a household word in the sport of polo (yes, as depicted in the shirt logo of Ralph Lauren), in part because a Jodhpur prince first introduced it to England. The royal families of Jodhpur, Jaipur, and other cities in the state of Rajasthan were versed in that sport. Jodhpur's distinctiveness derives not so much from excellence in the sport but in the devising of the pants worn by the contestants. They are tight fitting at the calf and below and loose at the hips. They are a variant of the traditional Indian garment known as the churidar. To this day, riding pants are known as jodhpurs, despite stretch fabrics now allowing for the hips to be close fitting as well.

Being on the edge of the Thar desert, we were subjected to dust storms, known locally as sandstorms. They varied in severity, but they all moved sand, in dust or grit form, from one location to another. As desert-dwelling kids, we knew the drill. One took occasional peeks at the horizon, not for lightning but for a wall of sand. An approaching one required immediate shelter from the suffocating sand and flying grit. We all carried handkerchiefs. If shelter was not close, these, tied around the mouth, had to do, combined with finding the highest elevation (did not want to be buried in the low-lying area). All this was known. Then, one day, I was 7 years old or so, and out and about with a friend, not very close to either of our homes. A cloudless sky appeared to be darkening in the distance. But it was just the sky, not closer to the ground, so this was not a sandstorm. Besides, it did not quite look like one. We sought counsel from the most wizened man (turns out it was a man; a wizened woman would have served the purpose; just saying) available. We could find just one man, wizened or otherwise. We concluded that he scored reasonably on the wizened scale. At that point, it was a choice to believe or not.

Are, wo kya hay? I asked, pointing to the dark sky in the distance. Rough translation: Say, what is that?

Wo tho tiddiyon ke baddal hay, he said, somewhat nonchalantly. Rough translation: That is a locust cloud (swarm).

We had heard of locusts. They were insects, albeit fairly large. But that was the extent of our collective knowledge. We were reassured by his seeming nonchalance. This was tempered by the possibility that, for all we knew, was because he was going to be safe because he knew what to do.

Wo kitna door hay? (How far is it?), I asked.

Bilkul pass he hay (It is quite close), he said.

I will spare you (and me) the rest of the vernacular. He went on to hazard a time much too short for us to run to either of our homes. He then explained that there was no cause for concern, the locusts simply wanted to feed on the crop nearby. He did suggest we remove ourselves away from the crop so as not to accidentally get run over, so to speak. This was sound advice that we took immediately to not test the avoidance radar of the critters.

We noted that some of the farmers employed one or both of two deterrents, smoke pots and drumbeats, neither of which appeared particularly effective, but I suppose they felt like they did something. Possibly the creatures were intended to be deflected to a nearby farm sans smoke and sound effects. Not unlike our deer fence in Chapel Hill: penetrable by thieves but not as penetrable as a neighboring property with no fence at all. The swarm came, and it blanketed the sky completely. Eclipsed the sun, as it were. They descended swiftly, like dive bombers, onto the crop and nothing else. In a short space of time there was no crop. No threat to us, whatsoever. Scary it still was. But, no dust.

References

1. West JJ, Cohen A, Dentener F, et al. What we breathe impacts our health: improving understanding of the link between air pollution and health. *Environ Sci Technol*. 2016;50 (10):4895–4904. https://doi.org/10.1021/acs.est.5b03827.
2. Syphard AD, Keeley JE. Location, timing and extent of wildfire vary by cause of ignition. *Int J Wildland Fire*. 2015;24(1):37. https://doi.org/10.1071/WF14024.
3. Ford B, Martin MV, Zelasky SE, et al. Future fire impacts on smoke concentrations, visibility, and health in the contiguous United States. *GeoHealth*. 2018;2(8):229–247. https://doi.org/10.1029/2018GH000144.
4. Kumar P, Robins A, Vardoulakis S, Britter R. A review of the characteristics of nanoparticles in the urban atmosphere and the prospects for developing regulatory controls. *Atmos Environ*. 2010;44(39):5035–5052.

5. Platt SM, El Haddad I, Pieber SM, et al. Gasoline cars produce more carbonaceous particulate matter than modern filter-equipped diesel cars. *Sci Rep*. 2017;7(1). https://doi.org/10.1038/s41598-017-03714-9.
6. Val S, Liousse C, Doumbia EHT, et al. Physico-chemical characterization of African urban aerosols (Bamako in Mali and Dakar in Senegal) and their toxic effects in human bronchial epithelial cells: description of a worrying situation. *Part Fibre Toxicol*. 2013;10 (1):10. https://doi.org/10.1186/1743-8977-10-10.
7. Schuetzle R, Schuetzle. *Catalyst and process for the production of diesel fuel from natural gas, natural gas liquids, or other gaseous feedstock*. U S Patent 9090831; 2015.
8. Bujak-Pietrek S, Mikołajczyk U, Kamińska I, Cieślak M, Szadkowska-Stańczyk I. Exposure to diesel exhaust fumes in the context of exposure to ultrafine particles. *Int J Occup Med Environ Health*. 2016;29(4):667–682. https://doi.org/10.13075/ijomeh.1896.00693.
9. United States Geological Survey. *Radioactive Elements in Coal and Fly Ash, USGS Factsheet*; 1997:163–197. Published. Accessed February 22, 2019 https://pubs.usgs.gov/fs/1997/fs163-97/FS-163-97.html.
10. Sui JC, Xu MH, Du YG, Liu Y, Yin GZ. Emission characteristics and elemental partitioning of submicron particulate matter during combustion of pulverised bituminous coal. *J Energy Inst*. 2007;80(1):22–28. https://doi.org/10.1179/174602207X174487.
11. Seames WS, Wendt JOL. The partitioning of radionuclides during coal combustion. *Adv Environ Res*. 2000;4(1):43–55. https://doi.org/10.1016/S1093-0191(00)00006-X.
12. Sinha CS, Sinha S, Joshi V. Energy use in the rural areas of India: setting up a rural energy data base. *Biomass Bioenergy*. 1998;14(5–6):489–503.

CHAPTER THREE

Ozone: Good high, bad nigh

Ozone is singled out for a chapter because of the important role it plays in the composition of particulate matter (PM). As discussed in Chapter 4, it is an ingredient, together with volatile organic chemicals (VOCs) and sunlight, in determining the composition of the organic layer on PM. The compounds formed, often in the ultrafine range in the coating (the substrate may be in the fine range), determine the toxicity.[1] Another reason for focusing on ozone is that its precursors are produced predominantly during the combustion that generates PM. These precursors include VOCs, carbon monoxide (CO), and oxides of nitrogen (NO_x) and come from two of the three major sources of PM, automotive exhaust and biomass combustion (the third source being electricity generation from coal).

$$NO_x + VOCs + sunlight \rightarrow O_3 \tag{3.1}$$

This reaction is limited by either of the two anthropogenic reactant concentrations. In any given sunlight exposure, the reaction may be limited by controlling either NO_x or VOCs. Means to control NO_x in diesel vehicles are discussed in Chapter 10. Interestingly, although, in the lean NO_x trap approach, the engine must be run rich periodically to regenerate the filter. Running rich means the deliberate production of hydrocarbons by running the engine at oxygen below stoichiometric quantities. Whereas this cleans the trap by reducing the trapped NO_x to nitrogen, unreacted hydrocarbons will inevitably add to the reaction above and produce ozone.

Formation and prevalence

Ozone is formed in both the stratosphere and the troposphere. The latter is at altitudes that are lower in the polar when compared with the tropical regions but an average of about 13 km. It contains most of the earth's atmosphere, especially the water molecules and is the region responsible for weather patterns. However, 90% of the ozone is in the stratosphere. The chapter title "Good high, bad nigh" succinctly states the role of ozone as beneficial to humans in the stratosphere but detrimental in the

Particulates Matter
https://doi.org/10.1016/B978-0-12-816904-9.00006-4

troposphere. The underlying mechanisms of impact comprise much of this chapter.

Stratospheric ozone is formed by the series of reactions known as the Chapman Mechanism:

$$O_2 + UV \rightarrow O + O \tag{3.2}$$

$$2O + 2O_2 + M \rightarrow 2O_3 + M \tag{3.3}$$

M is a third molecule present to catalyze the reaction and, characteristic of catalytic reactions, is regenerated (Fig. 3.1). The ultraviolet (UV) radiation strike is shown as the pink arrow on top. It breaks the oxygen molecule into the atomic constituents. Each atom collides with an oxygen molecule and the third molecule, whose purpose is to absorb the excess energy of the collision, thus allowing the reaction to proceed more rapidly. The third molecule is regenerated and available for further ozone production.

UV radiation is also responsible for the destruction of ozone, by the following reaction:

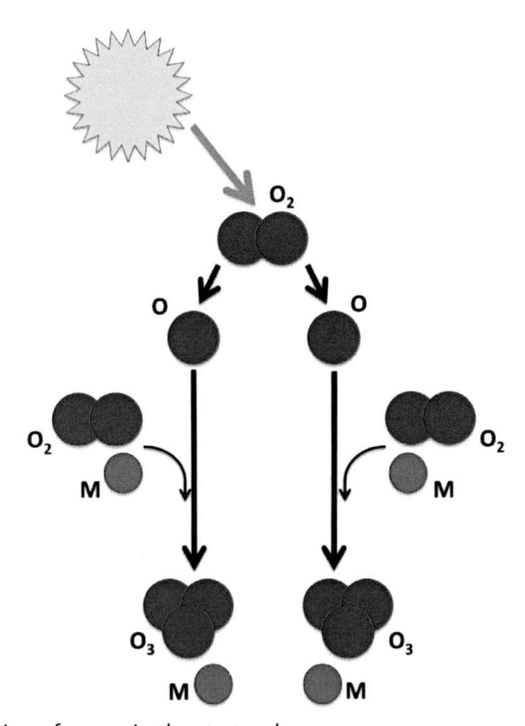

Fig. 3.1 Formation of ozone in the stratosphere.

$$O_3 + UV \rightarrow O_2 + O \quad (3.4)$$

$$O_3 + O \rightarrow 2O_2 \quad (3.5)$$

The equilibrium of the formation and destruction is disturbed by anthropogenic activity. Certain molecules break down ozone catalytically. This was proposed in a 1974 *Nature* paper by University of California, Irvine Professor Sherwood Rowland and postdoctoral fellow Mario Molina. They and the Swiss atmospheric scientist Paul Crutzen were awarded the 1995 Nobel Prize in Chemistry for this finding and for identifying the role of chlorofluorocarbons (CFCs) in breaking down ozone (see the close-up below). The reactions may be depicted pictorially as in Fig. 3.2.

The CFCs are broken down by UV radiation to produce a chlorine atom that collides with an ozone molecule to produce a chlorine monoxide (ClO) and an oxygen molecule. The ClO collides with an oxygen atom to produce an oxygen molecule, and the chlorine atom is released. Together with the original oxygen molecule created, there are now two oxygen molecules. The chlorine atom is now available to collide with another ozone molecule. This is a classic catalytic behavior, where the catalyst is not consumed. The significance of that in context is that the damaging impact of CFCs is long lived. CFCs have been used since the 1930s, primarily as refrigerants and as the propulsion agent in aerosol cans. Alternatives are now required in most countries to protect the ozone layer in the stratosphere, which is a key to limiting the effects of high-energy UV impacting the Earth.

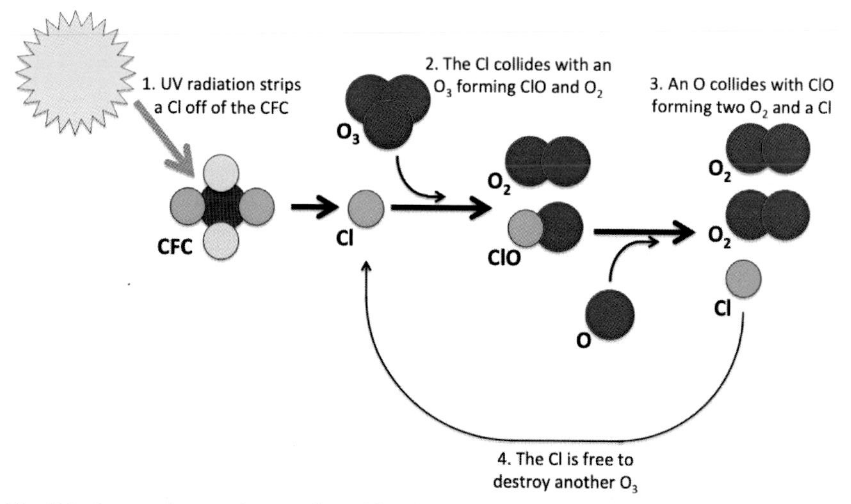

Fig. 3.2 Ozone destruction mediated by chlorine from chlorofluorocarbons.

The so-called ozone holes, areas where the ozone concentration is depleted but not necessarily eliminated, are partly responsible for the higher prevalence of skin cancer in parts of the world subject to them, such as Australia.[2] Whereas there are associations between the amount of UV light increased due to the ozone holes and skin cancer rates, there is very little direct causal evidence because of the 20- to 30-year lag time of skin cancer development from exposure. Even in Australia, researchers have found the ozone hole only partly responsible, with most of the cause coming from migration of fair skinned people to Australia and behavioral changes to the way in which Australians protect their skin.[3] The Victoria state government instituted a program to protect children. In 1981, a *Slip! Slop! Slap!* jingle was introduced, sung by a dancing Sid Seagull. It encouraged one to slip on a long-sleeved shirt, slop on sunscreen, and slap on a hat. Since then, rates of basal and squamous cell carcinoma have declined, but, curiously, melanoma incidence went up, as reported in a 1992 publication by Cedric Garland et al.[4] Importantly, the behavior is firmly imprinted in the psyche of the population, especially children, who were the targets of the cartoon seagull character in the messaging.

Close-up: Of salt, refrigeration, and cancer

Salt used to be critical for food preservation. It was a strategic commodity. People were partly paid in salt. Roman soldiers were reputedly paid in salt. Historians debate whether they were given coins enough to buy salt, or were actually paid in salt, or salt rations were added to payment, or were even paid at all! All of this notwithstanding, the word "salary" is believed to derive from salt. Now salt is merely useful and has not held strategic status for a few hundred years. The first step was the invention of canning, a response to a French government challenge, that is still with us today. The breakthrough was refrigeration. In simplest terms, it is concerned with reducing the temperature in a chamber to below ambient, to slow down bacterial-induced degradation. The systems became commercial when used in a closed cycle of compression and evaporative cooling. However, the chemicals used were either flammable or toxic. These included ammonia, sulfur dioxide, and propane.[5] Evaporative cooling has been used for hundreds of years to reduce the temperature of consumables.[6] To this day, in India, households will store water in a porous earthen jar, and modest evaporative cooling makes for a refreshing drink.

Refrigeration took a leap forward in the 1920s. Thomas Midgely and his team of chemists at Frigidaire resolved to create a refrigerant without the hazards of existing ones. They concluded that halogen-based organics were likely to have the necessary volatility and yet be stable. Their first compound

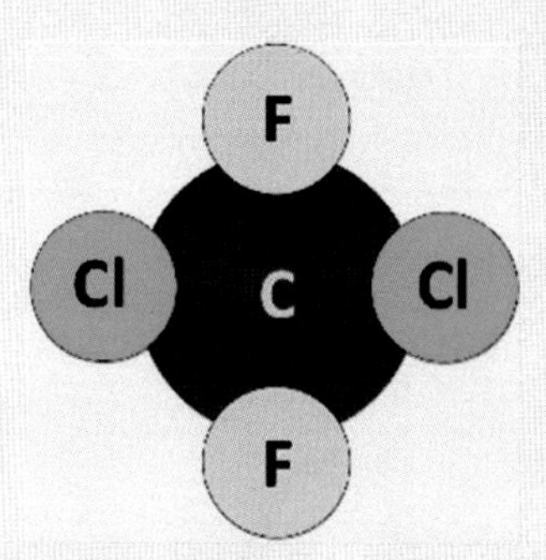

Fig. 3.3 Schematic of a CFC-12 molecule.

was dichlorodifluoromethane (CFC-12). Structurally, it looks like methane, CH_4, with the four H atoms replaced by two Cl and two F atoms, although the synthesis does not involve the methane molecule (Fig. 3.3).

They dubbed this class of compounds Freon, and it continues to be marketed under that name. CFC-12 is also known as Freon 12. This was recognized as an important advance in improving the well-being of people. Midgely was awarded the American Chemical Society's highest honor, the Priestley Medal, in 1941. Air conditioners, refrigerators, and to a lesser degree aerosol propulsion routinely used some versions of CFC. This started coming apart in the 1980s.

In 1972, Sherwood Rowland, a professor in the University of California, Irvine, heard a talk describing the work of James Lovelock in Great Britain. Lovelock had developed a sensitive means to detect small concentrations of airborne compounds. His results on trichlorofluoromethane (CFC-11) in the atmosphere suggested that all the CFC-11 ever made continued to exist. Rowland decided to investigate the possible implications of this long-lived behavior. He and Mario Molina, a postdoctoral fellow who had just joined his laboratory, studied available data and concluded that much of the CFC, stable in the troposphere, found its way into the stratosphere. They further concluded that under the right conditions, UV photons colliding with the CFC molecule would strip off an atom of chlorine. The chlorine atom would behave as follows:

$$Cl + O_3 \rightarrow ClO + O_2 \quad (3.6)$$

$$ClO + O \rightarrow Cl + O_2 \quad (3.7)$$

The chlorine atom is regenerated and is available to decompose yet another ozone molecule, in something like a chain reaction. The ClO intermediate compound is highly reactive. Rowland and Molina further concluded that other species could act in a similar fashion, including bromine and NO.

These findings were published in *Nature* and were regarded by the scientific community as important. However, experimental verification proved elusive. The species, some short lived, were in trace quantities, and measurement in the stratosphere required balloons in the main. Replication of conditions in the laboratory was equally daunting. The lack of experimental corroboration, and the extraordinary utility of the chemical, was the basis for industrial resistance to the elimination of CFCs. The chairman of the board of Dupont (which, at the time, had the patent on manufacture of CFCs) is quoted by Greenpeace[7] as referring to the Rowland work as "*a science fiction tale ... a load of rubbish ... utter nonsense.*" People took sides, but the scientific community remained convinced despite the lack of experimental confirmation.

Into this impasse stepped the British Antarctic Survey team, led by Joseph Farman, that had been making ground-based ozone measurements for decades. In 1985, they published a paper in *Nature*[8] concluding that the ozone layer in the stratosphere had been reduced by 40% at the end of the Southern Hemisphere winter, in September (cooler temperatures aid the reactions in question). This came to be known as the ozone "hole." Seen as a validation of the hypotheses of Rowland and Molina, world bodies acted in concert to target the replacement of CFC. The coordinated action among nations was required because the effects of emissions are not local.

The Antarctic hole is believed to affect Australia and New Zealand. As described in the body of this chapter, the Australians took steps to counter the effects of increased UV radiation. However, the 1992 letter to the editor by Garland et al.[4] draws attention to their belief that high SPF (sun protection factor) sunscreens may offer false security against melanomas and may even increase the incidence. They encourage the other portions of the Australian *Slip! Slop! Slap!* program, namely the long-sleeved clothing and hats. Heavy use of sunscreen predates this program in the state of Queensland that had the highest rate of melanoma in the world in 1987 when that finding was reported (reference 20 in the Garland paper; I am referring to it as such because the content is more like a research note than a letter to the editor). The dispute regarding the efficacy of sunscreens aside, little doubt exists regarding UV radiation being contributory to cancer of both types (carcinoma and melanoma) and that maintenance of the ozone layer in the stratosphere is a priority.

Tropospheric ozone is formed by the reaction of NO_x (which is some mixture of NO and NO_2) with CO or VOCs in the presence of sunlight. The process begins with the hydroxyl radical OH reacting with the hydrocarbons. In the case of CO, the reaction yields HOCO, a very unstable molecule, that immediately reacts with O_2 to produce a peroxy radical, HO_2.

$$OH + CO \rightarrow HOCO \quad (3.8)$$

$$HOCO + O_2 \rightarrow HO_2 + CO_2 \quad (3.9)$$

The peroxy radical oxidizes NO to NO_2. This is followed by NO_2 being broken down by the action of sunlight, primarily UVA wavelengths ($\lambda = 300$–400 nm), into NO and atomic oxygen. The oxygen atom reacts with O_2 to produce ozone.

$$HO_2 + NO \rightarrow OH + NO_2 \quad (3.10)$$

$$NO_2 + UV \rightarrow NO + O \quad (3.11)$$

$$O + O_2 \rightarrow O_3 \quad (3.12)$$

Given the presence of both NO_x and VOCs, either could be the limiting factor in ozone production for a given condition of ambient temperature and sunlight intensity. Researchers believe there are zones of NO_x and VOC limitation. A popular view is that NO_x limitation is the key. Certainly, when NO_x is below certain levels, tropospheric ozone levels remain low. In general, the warmer parts of the day in hot, dry areas with high solar intensity and concentrated vehicular use are the worst. Fig. 3.4 is a map of ozone intensity in the United States.[9] Hourly measurements are made at 81 CASTNET sites. Specifically, the image shows the fourth highest daily maximum 8-h average ozone concentrations for 2016. It certainly supports the rules of the thumb regarding the locations, where one could expect the worst tropospheric ozone. Later in this chapter, agricultural sources for ozone production will be discussed, with a focus on the new endeavor of cannabis cultivation. Much of this is in the western states, where ambient ozone could already be high.

The distribution of atmospheric ozone

The sun is essentially a black body radiator with a peak intensity of around 550 nm wavelength. Yet, terrestrial radiation has much of the spectrum missing below about 320 nm, at the UV end of the range. Scientists quickly deduced that ozone best fits the filtering characteristic.[10] Although present only at approximately 1 part per million (ppm), the layer effectively absorbs much of the harmful UV spectrum. The total ozone if concentrated

Fig. 3.4 Fourth highest daily maximum 8-h average of O_3 concentrations for 2016.[9]

into a layer would measure only about 3 mm in height. Ozone is measured in Dobson units (DU), and the 3 mm layer, although not present as a coherent layer per se, is equal to 300 DU. The unit is named after Gordon Dobson, at Oxford University, who made the first quantitative estimations of ozone in the stratosphere.

The ozone layer is present at about 25 km near the equator and gradually loses altitude at higher latitudes, down to about 15 km at the poles. The filtering out of portions of the UV spectrum is key to human endeavors, including plant life and human health. In the latter, the principal concern is UV-mediated skin cancer. This has already been discussed in the context of ozone hole formation due to anthropogenic activity. In summary, ozone high, in the stratosphere, serves a vital function in protecting the earth from high-intensity UV radiation. In contrast, ozone nigh, close to human endeavor, is harmful, especially at higher concentrations. Even low concentrations are known to be deleterious. As low as 0.2 ppm ozone caused significant airway inflammation from short-term exposure to diesel exhaust.[11]

Ozone and smog are often referred to synonymously, and this is wrong. Ozone is a component of smog, two of the other constituents being the precursor species that caused the nonnatural ozone to be formed in the first place. These are NO_x and VOCs, with the ozone forming by the reactions described above. The point being, in the absence of these anthropogenic species, ozone levels would be relatively benign and smog as we know would not form. The other typical constituents of smog, some of which give it the characteristic brownish color, are oxides of sulfur and PM. Two sources are responsible for most of these harmful constituents: industrial processes involving combustion and vehicles.

Health effects

Short term

Not debated is that ozone has deleterious effects on human health, especially affecting small children and the elderly. Short-term effects are proven, but long-term effects, especially on mortality, are subject to debate. Causal relationships are especially difficult to establish because the other constituents of polluted air will have independent effects on health as well. Some suggestions exist for coordinated action, such as ozone in the airways providing a stimulus for ultrafine PM to be absorbed. We do know, as described in Chapter 4, that the presence of ozone can increase the toxicity of PM by changing the character of the organic layer. Without being comprehensive, we will address most of these effects in this chapter.

An analysis of existing data has demonstrated that acute exposure to ozone results in *transient* impact on the respiratory system and *reversible* reductions in pulmonary function, together with an inflammatory response that can last 18–24 h.[12] Note the italics; these responses are not permanent, at least with the exposure durations tested, up to 6.6 h.

Fig. 3.5 shows experimental data on 30 healthy young adults from Brown et al. The significance of the levels of exposure is that, when the paper was published in 2008, the National Ambient Air Quality Standard for ozone was 0.075 ppm, and previous studies had examined levels above 0.08 ppm. The experiment was to measure lung function using the parameter forced expiratory volume in 1 s (FEV_1) at moderate levels of exercise. Of note is that the test at 0.06 ppm and the base case did not diverge until 4.6 h. The paper reports that there were minimal error bars to that point. They took this as evidence that the later divergence is statistically significant, even though small in physiologic terms. Note that the 0.08 ppm curve departs earlier, at 3 h. Reports from other studies show that the decrements in FEV_1 are up to 16% for ozone at 0.12 ppm, all at 6.6 h. Studies at this duration are used by the Environmental Protection Agency (EPA) to set their limits.

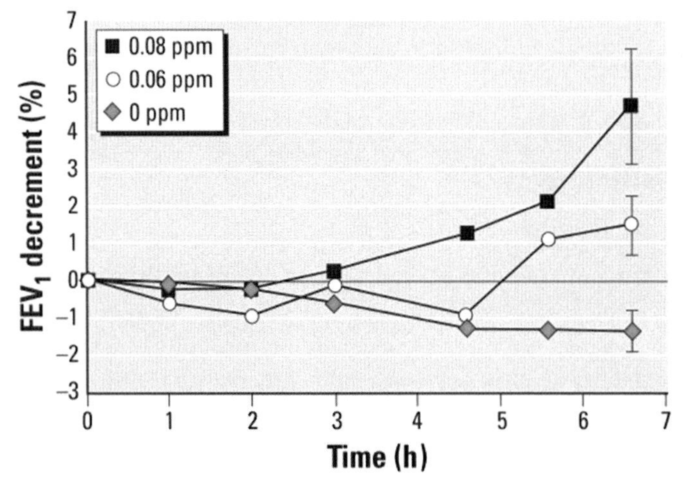

Fig. 3.5 FEV_1 as a function of exposure concentration and time.[12]

Ozone and PM

Ozone interacts with airborne particulates in two generic ways. One is a role played in the formation of organic coatings on PM, especially in the early stages of high reactivity due to the high surface area-to-volume ratio of ultrafine particles. This is covered in Chapter 4. The other, which will be discussed first, is the role in the physiology of humans, especially pulmonary function.

Epidemiologic studies on this issue are difficult because the ozone will be mixed with other airborne pollutants. Some studies with controlled variables were conducted using diesel exhaust and ozone at separate times. The rationale for sequential testing is that the peaks of each pollutant are usually found at different times of the day. Diesel exhaust is assumed to be the highest in the morning during rush hour (urban setting), whereas ozone could be expected to peak in the afternoon. A study of 16 healthy young adults was conducted while they were moderately exercising.[11] A 1 h exposure to diesel exhaust with PM concentration of 300 micrograms per cubic meter was followed 5 h later by a 2 h exposure to 0.2 ppm ozone. Induced sputum was collected 18 h after the ozone exposure. Statistically significant increases were observed in three different inflammation indicators. The investigators conclude that ozone present after a diesel exhaust insult results in greater levels of inflammation in the lung than the diesel exhaust itself. Because the particles in fresh diesel exhaust are largely expected to be in the ultrafine range, the extreme reactivity created by their size would make them inflammatory agents. In a separate study, removal of the ultrafine component eliminated cardiovascular damage.[13] Ozone insult a few hours later appears to have exacerbated the effect in the lung inflammation study. An epidemiologic study of 3535 children in Southern California compared the incidence of asthma between groups who played up to three sports outdoors with those who did not play. In high-ozone incident areas, significant increases were seen in the outdoor sport playing kids over those who did not. There was no difference in low-ozone areas.[14]

Long term

We have been discussing short-term exposure in the main, often described as seasonal exposure. Not much causal evidence links seasonal exposure to chronic obstructive pulmonary disease (COPD) mortality. But significant evidence may be found linking long-term ozone exposure and COPD mortality.[15–18] Large-scale studies typically must account for three principal constituents: $PM_{2.5}$, NO_x, and ozone. ($PM_{2.5}$ indicates particle diameter less than 2.5 μm.) The complications include the facts that NO_x is an ozone precursor and that $PM_{2.5}$ dominates the statistics. The authors of one such large-scale study, while acknowledging the interferences, are confident in the observation that long-term exposure to elevated ozone contributes to circulatory and respiratory mortality. They estimate that each 10 parts per billion of ozone causes a 3% and 12% increase in circulatory and respiratory mortality, respectively.[18]

Another study estimates that, in 2015, 8% of global COPD mortality could be attributed to elevated ozone, with China, India, and the United States leading the way. The associated mortality statistic is 254,000 COPD deaths.[15]

Ozone and climate change

This is another area, where "good high, bad nigh" applies. Stratospheric ozone has already been described as beneficial in blocking high-intensity UV radiation. Now, we must attend to the fact that the net effect of the radiation blockage is cooling on the Earth.[19] In other words, it has negative radiative forcing (RF), expressed as watts per meter squared (W m^2; see the box in Chapter 4 for a discussion of RF). However, tropospheric ozone has positive RF through interaction with terrestrial long wavelengths. Estimates of the greenhouse gas effect have relied on modeling. The IPCC AR5 report[20] shows that RF has increased by 0.40 ± 0.2 W m^2 since the Industrial Revolution. The associated global warming is estimated at 0.11°C.[21] However, all these estimates also recognize the short-lived feature of ozone in the troposphere because of its high reactivity. Furthermore, the synthesis also removes the independently deleterious precursors such as NO_x and VOCs.

Methane (CH_4) is not usually considered an ozone precursor, likely because it is relatively long lived. But, methane leakage from pipeline infrastructure and other parts of the natural gas delivery chain has become an important element in the global warming discussion. That debate has underlined the fact that fugitive methane emissions are also from municipal waste, animal flatulence, and coal mining.[22] Over 60% of methane emissions are considered anthropogenic. A study concludes that a 50% reduction in anthropogenic methane would be more effective in reducing tropospheric ozone than a 50% reduction in NO_x.[23] One caveat: the impact of NO_x reduction would be expected to be localized, whereas that of methane reduction would not. In some respects, the comparison is also artificial because the reduction measures apply to different industries. However, because methane is known to be a potent greenhouse gas by a direct mechanism, this additional suggestion of the indirect effect through ozone formation makes the target of reduction of fugitive methane even more important.

Cannabis: A new player in ozone production

The legal cultivation and distribution of cannabis in several US states has created an interesting problem with respect to ozone production. These plants generate VOCs in the form of terpenes. In the presence of sunlight and ambient NO_x, ozone is synthesized by the mechanism described above. Because this cultivation is essentially in violation of federal law, no federal legislation applies to emissions leading to the formation of ozone precursors that are unaccounted for in the National Emissions Inventory used by the EPA. The first state to allow the cannabis industry was Colorado, in 2014. Today, more than 600 indoor cultivation locations are within the city limits of Denver. Although grown in enclosed greenhouses, ambient air quality is affected because the gases produced are routinely pumped out of the building. Indoor cultivation is favored because the growing conditions can be altered at different stages of the plant's life cycle. Cannabis is native to the Indian subcontinent and needs high temperature and light intensity to achieve high yield. Denver greenhouses operate with light 24 h a day until just before flowering. That stage requires shorter "days." With further fine-tuning that would not be feasible in a field environment, the crop can have optimal yield and quality. Other controls to improve yield include elevating the CO_2 concentration to 1500 ppm, maintaining the temperature greater than 30°C, and controlling humidity.

Scientific studies in this area involve plants that do not have the growing advantages cited above. Nevertheless, the general principles apply, such as the relative proportions of the molecules produced. In a recent study,[24] four cannabis species were experimented upon. These were Critical Mass, Lemon Wheel, Elephant Purple, and Rockstar Kush (!). The terpenes in all the strains were dominated by β-myrcene (up to 60%), which is reputed to be the most dominant species in all cannabis-sourced terpenes. The other two were eucalyptol (up to 38%) and D-limonene (up to 10%).

β-myrcene: a variety of monoterpene, with the formula $C_{10}H_{16}$.

Eucalyptol: a type of terpenoid, with the formula $C_{10}H_{18}O$.

Limonene: a monoterpene, with the formula $C_{10}H_{16}$.

As one could conclude from the names, that these are naturally occurring in a host of other plants. Limonene is common in citrus, especially the rind; β-myrcene in ripe mangos; and eucalyptol, again as the name suggests, in eucalyptus leaves. As an aside, eucalyptus leaves are so abundant in oils that they are very flammable. A boon for boy scouts' campfire starting but not so much for forest fires. Forest fires, where the species is abundant, such as in Australia, are very dangerous: they hop from treetop to treetop.

The referenced study[24] estimates the ozone-forming potential (OFP) for three of the species they tested. The OFP was computed by the authors as follows:

$$\text{OFP} = \text{EC}\,(\mu g/g/h) \times \text{MIR}\,(\text{ozone}(g)/\text{VOC}(g)), \tag{3.13}$$

where EC is the emission capacity and is experimentally measured, and MIR is the maximum incremental reactivity, defined as the maximum ozone created per gram weight of terpene.[25] They plot the measured EC for three of the plants tested. There were insufficient data for Elephant Purple. The data (Fig. 3.6) are consistent with the earlier observation that the most abundant terpenes are the monoterpenes such as myrcene and terpenoids such as

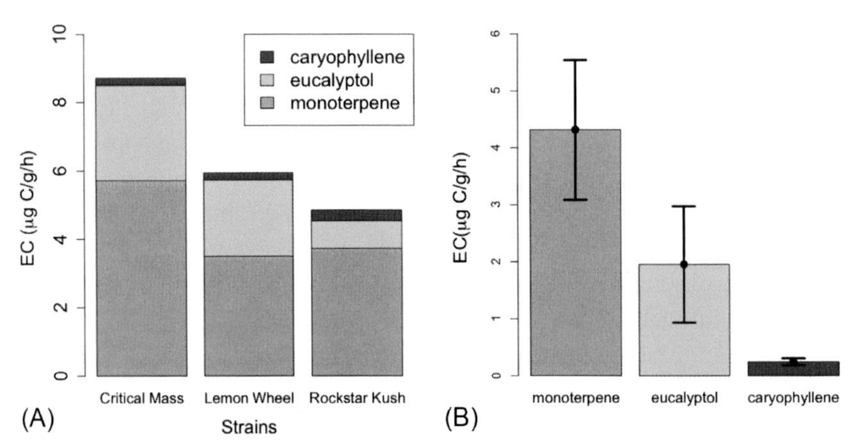

Fig. 3.6 Calculated emission capacities (ECs, expressed as μg/g/h) derived from measurements after 46 days of growth, normalized by dry leaf weight (g), and a standard temperature of 30°C for (A) Critical Mass, Lemon Wheel, and Rockstar Kush strains of *Cannabis* sp. and the variation of EC for (B) total monoterpenes, eucalyptol, and caryophyllene among the three strains.[24] *(With permission from Wang C-T, Wiedinmyer C, Ashworth K, Harley PC, Ortega J, Vizuete W. Leaf enclosure measurements for determining volatile organic compound emission capacity from* Cannabis *spp. Atmos Environ 2019;199:80–87. doi:10.1016/j.atmosenv.2018.10.049.)*

eucalyptol. The variability in production from different species is also evident in panel A.

The OFP computed from these data also demonstrates variability, with eucalyptol in Critical Mass being the heaviest producer at about 12 μg/g/h (data not shown). The takeaway is that the ozone production is species specific and needs to be estimated for ameliorative action to be taken. The authors did estimate an expected overall impact in the Denver area. Each of the 600 operations is permitted to carry 10,000 plants. With an assumed biomass of 1 kg/plant, an EC of 8.7 μg/g/hour, and assuming that all emissions would be released into the surroundings of the greenhouses, they compute annual terpene emissions to be 520 tonnes. They state that this is more than double the total biogenic VOC emissions from all other sources in the area. The ozone created by the 520 tonnes of terpenes is estimated to be 2100 tonnes per year. Especially, during the hot months, this could be highly material to the total ozone level. Note also in Fig. 3.4 that Colorado has some of the higher ozone loadings in the nation. One would expect ambient conditions to be worse in the other important cannabis-growing area, California. The authors also compute the production of PM comprising secondary organic aerosols as a result of these emissions (see Chapter 4 for mechanisms of secondary organic aerosol formation) to be 131 tonnes per year. This too adds to the pollution burden, in a different way. More studies such as this, likely informed by the expected varieties to be grown in each area, will be necessary to truly understand the degree to which this new cultivar is affecting ambient air pollution. Future studies ought also to include variances in terpene production over the life cycle, with particular attention to greater production in the flowering phase. The uncertainties in the cited study include:

- The MIR values are default estimates in some cases.
- The ECs in commercial cultivation will almost certainly be greater than in the laboratory setting due to the optimization means used, such as CO_2 addition and manipulation of light.
- The assumption of 1 kg per plant may also be inaccurate, most likely on the low side.

The terpenes will not produce ozone without NO_x. Much of the NO_x in these settings is from vehicles and unlikely to be the limiting factor in urban settings (some conventional cannabis growing fields in California are known to be sufficiently remote as to be NO_x limited). The solution lies, therefore, in addressing the terpene production. Later in this book, engineered solutions that could ameliorate this problem are discussed.

References

1. Delfino RJ, Staimer N, Tjoa T, et al. Associations of primary and secondary organic aerosols with airway and systemic inflammation in an elderly panel cohort. *Epidemiology*. 2010;(6):892–902. https://doi.org/10.1097/EDE.0b013e3181f20e6c.
2. Henriksen T, Dahlback A, Larsen SHH, Moan J. Ultraviolet-radiation and skin cancer. Effect of an ozone layer depletion. *Photochem Photobiol*. 1990;51(5):579–582. https://doi.org/10.1111/j.1751-1097.1990.tb01968.x.
3. Sánchez CF. The relationship between the ozone layer and skin cancer. *Rev Med Chil*. 2006;134(9):1185–1190. https://doi.org/10.4067/S0034-98872006000900015.
4. Garland CF, Garland FC, Gorham ED. Could sunscreens increase melanoma risk? *Am J Public Health*. 1992;82(4):614–615. https://doi.org/10.2105/ajph.82.4.614.
5. United States Environmental Protection Agency. *Refrigerant Safety [Overviews and Factsheets]*; 2014, November 5. Retrieved July 9, 2019, from US EPA website https://www.epa.gov/snap/refrigerant-safety.
6. Watt JR. History of evaporative cooling. In: Watt JR, ed. *Evaporative Air Conditioning Handbook*. New York: Chapman and Hall; 1986:5–11. https://doi.org/10.1007/978-1-4613-2259-7_2.
7. Greenpeace Position Paper. *DuPont: A Case Study in the 3D Corporate Strategy*; 2004, March 15. Retrieved July 9, 2019, from https://web.archive.org/web/20120406093303/http://archive.greenpeace.org/ozone/greenfreeze/moral97/6dupont.html.
8. Farman JC, Gardiner BG, Shanklin JD. Large losses of total ozone in Antarctica reveal seasonal ClO_x/NO_x interaction. *Nature*. 1985;315(6016):207–210. Bibcode:1985Natur.315..207F https://doi.org/10.1038/315207a0.
9. EPA. *EPA Clean Air Status and Trends Network (CASTNET) Annual Report*; 2016. chapter 2 https://www3.epa.gov/castnet/docs/CASTNET2016/AR2016-main.htm#chapter2top. retrieved August 4, 2020.
10. McElroy CT, Fogal PF. Ozone: from discovery to protection. *Atmos Ocean*. 2008;46(1):1–13. https://doi.org/10.3137/ao.460101.
11. Bosson J, Pourazar J, Forsberg B, Ädelroth E, Sandström T, Blomberg A. Ozone enhances the airway inflammation initiated by diesel exhaust. *Respir Med*. 2007;101(6):1140–1146. https://doi.org/10.1016/j.rmed.2006.11.010.
12. Brown JS, Bateson TF, McDonnell WF. Effects of exposure to 0.06 ppm ozone on FEV1 in humans: a secondary analysis of existing data. *Environ Health Perspect*. 2008;116(8):1023–1026. https://doi.org/10.1289/ehp.11396.
13. Mills NL, Miller MR, Lucking AJ, et al. Combustion-derived nanoparticulate induces the adverse vascular effects of diesel exhaust inhalation. *Eur Heart J*. 2011;32(21):2660–2671. https://doi.org/10.1093/eurheartj/ehr195.
14. McConnell R, Berhane K, Gilliland F, et al. Asthma in exercising children exposed to ozone: a cohort study. *Lancet*. 2002;359(9304):386–391. https://doi.org/10.1016/S0140-6736(02)07597-9.
15. Cohen AJ, Brauer M, Burnett R, et al. Estimates and 25-year trends of the global burden of disease attributable to ambient air pollution: an analysis of data from the global burden of diseases study 2015. *Lancet*. 2017;389(10082):1907–1918. https://doi.org/10.1016/S0140-6736(17)30505-6.
16. Fann N, Lamson AD, Anenberg SC, Wesson K, Risley D, Hubbell BJ. Estimating the national public health burden associated with exposure to ambient PM2.5 and ozone. *Risk Anal*. 2012;32(1):81 95. https://doi.org/10.1111/j.1539-6924.2011.01630.x.
17. Tager IB, Balmes J, Lurmann F, Ngo L, Alcorn S, Künzli N. Chronic exposure to ambient ozone and lung function in young adults. *Epidemiology*. 2005;16(6):751–759. https://doi.org/10.1097/01.ede.0000183166.68809.b0.

18. Turner MC, Jerrett M, Pope CA, Krewski D, Gapstur SM, Diver WR. Long-term ozone exposure and mortality in a large prospective study. *Am J Respir Crit Care Med.* 2016;193(10):1134–1142. https://doi.org/10.1164/rccm.201508-1633OC.
19. Forster P, Ramaswamy V, Artaxo P, et al. Changes in atmospheric constituents and in radiative forcing. Chapter 2. In: *Climate Change 2007: The Physical Science Basis. Contribution of Working Group I to the Fourth Assessment Report of the Intergovernmental Panel on Climate Change, Intergovernmental Panel on Climate Change, Ed.* Cambridge University Press; 2007.
20. Field CB, Barros VR, Dokken DJ, et al, Intergovernmental Panel on Climate Change (IPCC). Climate change 2014: impacts, adaptation, and vulnerability. Part A: global and sectoral aspects. In: *Contribution of Working Group II to the Fifth Assessment Report of the Intergovernmental Panel on Climate Change.* New York: Cambridge University Press; 2014.
21. Shindell D, Faluvegi G, Lacis A, Hansen J, Ruedy R, Aguilar E. Role of tropospheric ozone increases in 20th-century climate change. *J Geophys Res.* 2006;111. https://doi.org/10.1029/2005JD006348.
22. Rao V, Knight R. *Sustainable Shale Oil and Gas: Analytical Chemistry, Geochemistry and Biochemistry Methods.* Cambridge, MA: Elsevier; 2016.
23. AM Fiore AM, Jacob DJ, Field BD, Streets DG, Fernandes SD, Jang C. Linking ozone pollution and climate change: the case for controlling methane. *Geophys Res Lett.* 2002;29(19). https://doi.org/10.1029/2002GL015601. 25-1-25-4.
24. Wang C-T, Wiedinmyer C, Ashworth K, Harley PC, Ortega J, Vizuete W. Leaf enclosure measurements for determining volatile organic compound emission capacity from *Cannabis* spp. *Atmos Environ.* 2019;199:80–87. https://doi.org/10.1016/j.atmosenv.2018.10.049.
25. California Air Resources Board. *Tables of Maximum Incremental Reactivity (MIR) Values*; 2010. Retrieved from https://ww3.arb.ca.gov/regact/2009/mir2009/mir2009.htm.

PART TWO

Impacts of differences in particle size

Part One dealt with sources. This part is about mechanisms of PM formation and the actions of particles on organs. Regardless of source, all PM is formed in the same way: the formation of a nucleus and the growth of the particle from the deposition of organic matter. The organic matter may be coproduced or in the ambient air sourced from elsewhere.

The nucleus largely comprises elemental carbon and organic droplets. The three principal sources—coal, wood, and hydrocarbon—have differing proportions of coproduced organics. Burn conditions also determine organic content. Final particle size determines toxicity and mortality, attributable primarily to $PM_{2.5}$, that is, particles smaller than 2.5 μm.

Health effects are described in some depth for both morbidity and mortality. Smaller particles bypass the mucus traps in the upper respiratory system and enter the alveoli in the lung. The smallest particles, under 0.1 μm, can cross cell boundaries, making them the most toxic. All major organs can be affected, including the brain. Both long- and short- term impacts of PM on health are discussed.

CHAPTER FOUR

The importance of being small (*with apologies to Oscar Wilde*)

Conventional particulate matter (PM_{10}, $PM_{2.5}$) and ultrafine

Size certainly matters. But in this realm, smaller is more powerful, not necessarily in a good way. In fact, decidedly not in a good way when it comes to toxicity. In every step along the way from inception of the particle nucleus to growth and transport and finally to action on human organs, size plays a critical part, and the smallest particles, less than 0.1 μm (μm), are the most effective in each of those reactions. But even the moderately smaller particles, less than 2.5 μm, are significantly more toxic than larger particles. In this chapter, we describe the various roles played by size in the reactions involved in the entire life cycle of these particles.

Particulate matter (PM) ranges in size from a few nanometers (nm) to 100 μm. That covers four orders of magnitude. The range is another few orders of magnitude greater when one considers the surface area of the particles instead of diameter. Surface area is the key parameter in the chemical reactions affecting particles. Not surprisingly, size does matter in the impact on human endeavor. Except, in the PM world, the littlest guys kick serious sand into the eyes of the large ones.

PM is a grab bag of species ranging from liquid to solid, and inorganic to organic, with the latter also excursing into bioaerosols. In this book, we largely limit ourselves to species that originate from the combustion of carbonaceous material, as described in Chapter 2. That is because most particulates concerned with human health are so derived. This is not to say that oxides of various description, comprising the milieu known as dust, are not matters of health concern. Those species, primarily some combination of oxides of silicon and aluminum, are sourced naturally. Silica (oxide of silicon) is prominent in dust originating in deserts. Together with alumina (oxide of aluminum), they are the principal ingredients of clay, a ubiquitous

Particulates Matter
https://doi.org/10.1016/B978-0-12-816904-9.00009-X

material in soil. When these materials are mined and used commercially, such as silica in oil and gas operations and cement manufacture, the dust created in the handling could well be classified as anthropogenic in character. Silicosis is a well-known disease from occupational exposure. Asbestos is a naturally occurring silicate, most commonly found in association with the attractive mineral serpentine that is the "state rock" of California (seriously, state rock). Industrial use has serious health effects, including mesothelioma. Exposure to carbon particles during coal mining causes black lung disease, a well-studied matter that is not addressed here because intervention methods are known, whether employed across the board or not. In the United States, the Occupational Safety and Health Administration is responsible for regulations to protect workers.

Particulate matter definitions and nomenclature

PM_{10}: Smaller than 10 μm, passing through a filter with a cutoff at 10 μm in aerodynamic diameter to a 50% efficiency.

$PM_{2.5}$: Smaller than 2.5 μm, passing through a filter with a cutoff at 2.5 μm in aerodynamic diameter to a 50% efficiency.

$PM_{0.1}$: Smaller than 0.1 μm (100 nm), passing through a filter cutoff at 0.1 μm.

Coarse particles: Larger than 2.5 μm and smaller than 10 μm.

Fine particles: Smaller than 2.5 μm but larger than 0.1 μm.

Ultrafine (UF) particles: Smaller than 0.1 μm but larger than 0.05 μm (50 nm).

Nanoparticles: Smaller than 50 nm. The air quality community tends to follow this nomenclature. The materials science industry defines nanoparticles as simply below 100 nm. The reason for this choice is that, for most materials studied by the industry, particles below this diameter tend to have dramatically different physical properties than their larger counterparts. This is due to the much larger surface-to-volume ratio, an aspect that also comes into play in aerosol reactions.

A point of note: Each of these categories specifies the *maximum* size. The particle distribution will go down to the next category.

How particulates form

The discussion will be limited to formation in the energy sector because that sector, comprising the production and use of fuels, is responsible for 85% of airborne particulate matter.[1] Other sources will be referred to, but the energy related ones will be examined in some detail. They all entail combustion of a fuel. The three principal fuels are coal, diesel, and wood, with a further nod to postharvest crop residue such as rice stalks and bagasse (sugarcane fibers after sugar extraction).

An interesting source, although relatively minor in scale, is flaring (combustion) of natural gas associated with oil production. As we will not discuss this topic again, it merits some description. When oil is produced, in many instances, the gas associated with it is in too small a quantity to merit a pipeline, so it is flared. In general, the lighter the oil, the greater the amount of associated gas. This is a result of the mechanism of oil and gas formation from organic matter under high pressure and temperature in the earth. Oil is less thermally mature than gas. Light oil—that is, oil with relatively smaller molecules—is comparatively more thermally mature. Accordingly, it often has the most mature gas species, methane, in association. Inevitably, it also has the intermediate size molecules, such as propane and butane. In that case, the gas is deemed to be "wet."

A 2015 estimate has annual flaring of 5.4 trillion cubic feet of gas per annum. In many instances, the control of oxygen stoichiometry is not tight, resulting in unburnt hydrocarbons in addition to the soot. This is in part because wet gas associated with light oil, such as shale oil, will contribute to a richer (oxygen starved) burn. In everyday life, when one sees a yellow flame, it is oxygen starved. Candles rely on this property to shed light. A blue flame, on the other hand, has a more complete burn. Burners for cooking most advantageously have a blue flame. Whereas the overall contribution to PM worldwide from this flaring source is small (3%), in the Arctic, it is over 40%.[2] Much of this is black carbon, resulting in acceleration of ice melting due to solar heat absorption by the carbon deposited on the ice surface.

All particulates share a common route to formation: nucleation, followed by growth. Nucleation can be either homogeneous or heterogeneous. In homogeneous nucleation, molecules simply combine to form the nucleus. This is not as energetically favored as heterogeneous nucleation, in which the molecules attach to an existing particle. In principle, the nucleating agent could also be an energetically favored surface in the discharge equipment, such as the walls of a diesel engine exhaust system. This mechanism is particularly common in reactions involving changes of state. A common example is cloud formation, where water vapor condenses to liquid when a certain vapor saturation is reached. This can be accelerated by the presence of small solid particles, such as silver iodide or potassium iodide crystals. This process, known as cloud seeding is debated regarding its efficacy, but the science is sound. In general, the smaller the seed particle, the more effective the ability to attract molecules.

Fig. 4.1 schematically describes the progression of particle formation and growth in all combustion processes. The initial nucleus is small, 10–40 nm in diameter. The material is most commonly carbon but can also be liquid droplets. In general, the heterogeneously nucleated particles will be somewhat smaller than those nucleated homogeneously because of the lower energy barrier to formation.

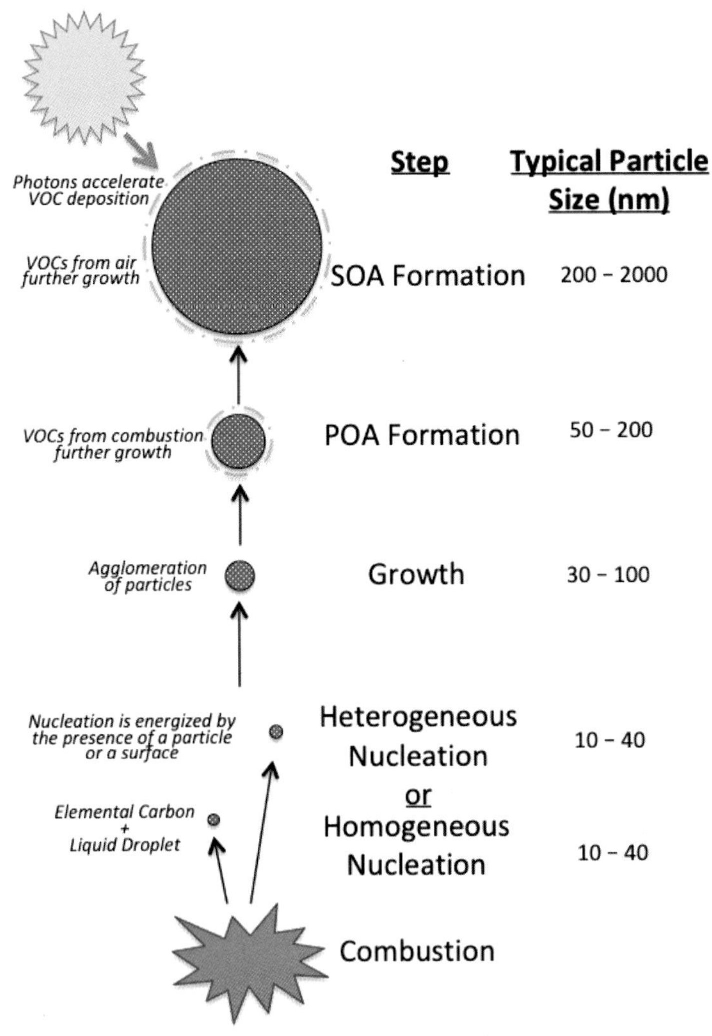

Fig. 4.1 Stages of formation and growth of particulate matter. SOA = secondary organic aerosol, POA = primary organic aerosol.

The next step is growth, primarily through agglomeration of particles—the accumulation mode. After some growth, particles can acquire a coating of organic material. When the source for the organics is volatile organic chemicals (VOCs) originating from the material being combusted (e.g., unburned diesel fuel when the engine is running rich), the resulting particle is a primary organic aerosol. When the core of the particle is carbon, as is often the case, the term is primary organic carbon that is discussed later in the chapter.

Volatile organics: Good nigh, bad high

In much of this book, VOCs are villains in black hats. Their role in increasing the toxicity of PM is noted in this chapter and expanded upon later in the book. To set the record straight, volatile organics are heroes in white hats in their preferred settings. Role reversals occur when they leave home. The terpenes that protect the flowering plant contribute to ozone production in the surrounding environment if oxides of nitrogen (NO_x) are present (Chapter 3). They and other organics are involved in the formation of toxic organic coatings on PM in the presence of sunlight.

Ethylene is a hydrocarbon and the smallest alkene, with the formula C_2H_4. It forms the backbone of much of what we know as plastics. It was identified in 1935 as a plant hormone involved in fruit ripening. This volatile organic is routinely used in commerce for fruit ripening. The home remedy equivalent is placing a banana in a paper bag with the fruit to be ripened. The ethylene released by the ripening banana accelerates the ripening of the mango (my preferred fruit in season). We veterans of this technique have noticed on occasion that nothing much happens. A recent study in *Nature Plants*[3] explains why. A genetic switch comes into play at some stage of ripening. Prior to that, ethylene fumigation will be ineffective at best. There, in a nutshell, is my "nothing happens" scenario with the banana in the bag. Too soon, apparently. In my defense, when a mango, destined for yellowness, is green, hazarding a guess as to stage of ripening is hard. Not to mention, who knew about this genetic switch deal? Perhaps mangos, and peaches, ought to come with the modern equivalent of the pop-up thermal indicators in turkeys being baked. The Indians ought to be inventing a color-changing indicator of the beginnings of ethylene production instead of chasing landing craft around the moon (being of Indian origin gives me the right to be snide; I think). Incidentally, not all fruit responds to the banana-in-bag ripening; pineapples, grapes, and most berries are out. While on the subject, do not store tomatoes in the refrigerator. According to a 2016 study published in *PNAS*,[4] many of the genes responsible for flavor are irretrievably lost.

Terpenes are the most abundant VOCs produced by plant matter. The aroma from rosemary and mint, to name just two herbs, results from the volatilization of terpenes off the leaflets. In many species, such as cannabis, they are produced more during flowering. The function appears to be toward off predators. Some of the chemicals secreted attract beneficial species such as predators for the harmful ones. Rather than pursuing a litany of beneficial functions, it would suffice to assert that VOCs such as terpenes have a vital role in the survival of plant matter. The fact that they have a deleterious effect on ozone production and PM toxicity merely points to the need for remedial means for controlling the release into the atmosphere from cultivated plants. To sum up, the role of plant-sourced VOCs: they are good nigh and bad high.

The next step in the progression of the particle is further growth and deposition of organic molecules derived from ambient VOCs and mediated by photochemical action from sunlight-sourced photons. Ambient VOCs are largely anthropogenic but can also come from vegetation, such as terpenes from pine trees. Several species of commercially grown plants emit VOCs, including, famously, terpenes from cannabis cultivation. The differences in toxicity between primary and secondary organic aerosols are important and the subject of discussion in several other chapters.

Fig. 4.1 also shows typical ranges for particle size in each phase of the cycle. These can be variable and are primarily intended here to be illustrative of the orders of magnitude involved from inception to formation of particles as large as 10 μm (PM_{10}).

Champagne bubbles: All about the glass

Bubbles are the essence of champagne. Without them, it is just chardonnay (or pinot noir or a combination thereof). Many believe that the best sparklers have the smallest bubbles. Hmmm. Once we go through the physics, you be the judge. The wine as made will have carbon dioxide dissolved under pressure. Therefore the glass is thicker walled, and the cork has a retaining cage. One should be careful when removing the cage; I have had them pop right off on me. In company, that is not recommended, especially if the bottle is carelessly pointed. When poured into the glass, the first rush of bubbles will occur in any kind of glass. But the more completely dissolved gas needs spots in the glass upon which each bubble attaches and is then released. These are known as nucleation sites.

Blown crystal glasses provide the best nucleation sites. This is because the bottom of the glass joining the stem to the fluted part is full of little imperfections, basically minute cracks. The gas bubbles nucleate on these sites. As each bubble releases, another one forms on the site, and soon you have a steady stream. (You may have noticed that the bubbles *always* originate from the bottom, never from the sides.) So, does higher quality wine produce smaller bubbles? More than likely yes. But it still is all about the glass. Go ahead, pour some good stuff (loosely defined by how much you paid for it) into a water tumbler or any pressed glass wine flute. Then pour it into blown crystal. Does not even have to be a classic flute, any blown crystal glass will do for the demonstration. The difference in the bubble plume ought to be dramatic, especially after a few minutes when the pressed glass liquid loses steam. If it is not, your crystal glass may well be crystal (over 20% lead oxide in the silica), but it is decidedly not blown. Here also is some advice for the soon to be wedded. Forget about high-end crystal red wine glasses; just any big ones with as thin a rim as possible will do. But get blown crystal champagne flutes.

Particulates from coal combustion

Of the three sources we examined in this book, coal combustion is the most different. Coal is classified as a hydrocarbon, but it is seriously hydrogen challenged. Coal is formed by vegetative matter getting buried under layers of sediment. The overburden (the accumulated layers of sediment deposited over millions of years) exerts pressure, and temperature always increases with geological depth. Although temperature gradients vary, they tend to average 25–30 °C per kilometer of depth. The combination of pressure and temperature converts the vegetation to coal. Methane may also be found in association, but the mechanism for formation is largely biogenic (due to bacterial action). Methane formation is one reason why coal mining is dangerous. It is also the source of coal bed methane, a fossil fuel source.

At shallower depths, the coal is less mature, with lower fixed carbon and a higher proportion of moisture and clay-like minerals entrained in the coal seam. This is low-grade coal, and a well-known variant has the name lignite. Worldwide, nearly half of the coal reserves are low grade.

The minerals in coal are prominent in our discussion of emissions. The combustion process causes these species to be present in the ash. The minerals comprise, in the main, oxides of silicon, aluminum, and a few other elements. Being oxides, they do not combust, but they do melt at combustion temperature. They then recrystallize, sometimes into hollow spheres known as cenospheres, often filled with gas, with specific gravities significantly less than one. The industrial practice of storing fly ash in ponds is essentially fraught because at that specific gravity, this component of the fly ash will not settle to the bottom as intended.

Some components of the ash, such as oxides of sodium (Na), potassium (K), and arsenic (As), are highly volatile and can volatilize directly. But for the higher melting components such as oxides of calcium (Ca), silicon (Si), and magnesium (Mg), the volatilization occurs through a reduction of the liquid phase by the carbon monoxide (CO) produced at the char surface. The reaction in the case of SiO_2 is

$$SiO_2\,(\text{liquid}) + CO \rightarrow SiO\,(\text{vapor}) + CO_2 \quad (4.1)$$

Typical pulverized coal combustion temperatures exceed 1200°C. At these temperatures, the surface of the coal particles can get hot enough to yield a local reducing atmosphere (oxygen starved) that promotes the formation of CO, as opposed to oxidizing all the way to CO_2. This is the CO that participates in the vaporization of ceramic oxides such as SiO_2, as shown in the equation earlier.

These vapor species condense to nanometer size particles that in turn act as nucleating agents for all manner of species, including oxides still in the vapor state. In the main, the final particles produced are in the range of fine and ultrafine. Because one mechanism for the formation is vaporization, reducing flame temperatures will disfavor formation of these particles. The lower temperatures will also reduce NO_x formation, thus positively correlating the amelioration of NO and NO_2, the principal constituents of NO_x. However, under these conditions, the proportion of unburnt carbon in the fly ash increases due to the oxygen starvation. When the proportion of unburnt carbon is greater than about 6%, the fly ash is not suitable for beneficial use in concrete. But processing methods exist for bringing down the carbon to acceptable levels prior to use.[5]

Most industrial processes for combustion of coal use one of two means to capture the fly ash: bag houses containing fabric bags that act as filters, and electrostatic precipitation in which the particles are charged and then sent into an electrostatic field, where upon they are diverted for collection. Both methods are less effective on UF particles. Consequently, UF particle count and chemistry are an important area of study with respect to impact on human health. Such particles, escaping to the atmosphere, are more prone to adsorbing organic chemicals in the air, rendering them different and likely more toxic than the original. A small body of literature posits the notion that trace metals in coal fuse at combustion temperatures and form nucleating sites for species.[6] Although relatively small in total proportion, the actual numbers of particles could be significant. About 1 million particles of size 0.1 μm equal a single 10-μm particle on a volume basis. This fact continues as a refrain in this book: UF particles, although small in mass, are potentially harmful out of proportion to their mass. World Health Organization guidelines continue to be based on mass (micrograms per cubic meter) and may well be out of step with physiological impact.

Particulates from use of fuel

Of the major sources of PM, coal for electricity is the only one where the purpose is not solely direct use. Coal may be used for another purpose, the production of synthesis gas (syngas) to make a host of petrochemicals for which syngas is the raw material. As discussed in Chapter 2, syngas from coal is an intermediate even for "clean coal" electricity. We established that syngas was, by and large, not a PM source.

Some generalizations about particle formation may be drawn regardless of type of biomass being combusted. Over 90% of the carbon is oxidized to CO_2 or CO, depending on combustion conditions. Usually, less than 5% is released as carbon in particulate form. Most particles emitted are in the form of organic carbon. Although the nomenclature remains challenging, organic carbon is the carbon in the organic molecules. It could have been sourced by carbon particles that had organic molecules deposited on them. If the organic molecules are coproduced, then the particle will have a certain characteristic. These particles are called primary organic carbon. However, they could also be manufactured in the atmosphere by the action of VOCs from a different source and mediated by photochemical action. In this case, they are secondary organic carbon. A clear distinction has not been drawn, in terms of toxicology, between these two types,[7] likely because of the difficulty in separating out the variable of aging of the particles. Studies to date have joined together the species, and they could well have different effects on the human body. Our discussion of diesel exhaust-related aerosols describes one such study. Even if the aerosol is formed immediately following combustion, it may undergo further modification, usually deposition of further organics in the atmosphere. The smaller the original emitted particle, the more likely the secondary action, due to smaller particles having a higher surface-to-volume ratio and, therefore, are more energetically favored for deposition. In other words, these early assemblages act as nucleating agents for the growth mode. The original nucleating particle coming out of the combustion is usually either black carbon, brown carbon, condensed oxide vapor as described above, or an early assemblage of polycyclic aromatic hydrocarbons, formed likely by homogeneous nucleation. In the case of coal, and possibly some other sources, vaporized and later condensed metals can be the nucleating agent. That is, the original nucleating agent may be a solid or a liquid.

Black carbon appears not to have a standard definition. In many instances the definition centers around the light-absorbing character. The absorption cross-section for short-wave infrared radiation is significantly higher for this material than for other aerosol species. It is often described as "graphitic" or "elemental" carbon. In fact, the particle will almost always have some associated hydrogen and oxygen and possibly nitrogen, albeit in very small proportions. One source estimates the formula to be C_8H that, on a weight basis, makes it a severely hydrogen-challenged hydrocarbon—not unlike coal devoid of the mineral matter. But, in most cases, it will act like elemental carbon in the character of absorbing infrared and visible light, especially the former.

Brown carbon has entered the environmental consciousness recently and, therefore, deserves special attention. It too is difficult to define. The most available definition is an organic carbon capable of absorbing light. It is characteristically light brown in color because that is the portion of the spectrum reflecting off the surface of the particle.

Recent attention has been occasioned by the discovery that it is a significant contributor to climate change. Modeling at Argonne National Laboratory estimates that up to 19% of warming is due to brown carbon. They use this result to emphasize the need to include this species in radiative-forcing models.[8]

Radiative forcing

The Intergovernmental Panel on Climate Change (IPCC) defines radiative forcing as *"a measure of the influence a factor has in altering the balance of incoming and outgoing energy in the Earth-atmosphere system and is an index of the importance of the factor as a potential climate change mechanism."*

A portion of incoming solar radiation is absorbed in the atmosphere above the Earth, and a portion is radiated back out into space. Anthropogenic species can affect this balance. If there is a net absorption, then this is positive radiative forcing and causes a net heating effect on the Earth. It is expressed in watts per meter squared ($W\ m^{-2}$) and attributed to the species in question. Conversely, radiative cooling is accomplished when the balance goes in the other direction. Curiously, brown carbon appears to do both: positive radiative forcing in the upper atmosphere and negative cooling forcing on the surface (Fig. 4.2).

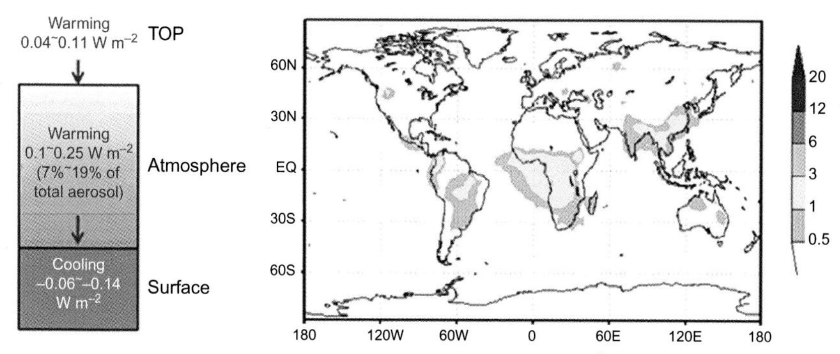

Fig. 4.2 Atmospheric warming and surface cooling (in W m^{-2}) due to light absorption by organic brown carbon aerosol. Note the geographical distribution of higher concentrations of brown carbon.[9] *(Source: Argonne National Laboratory.)*

Many investigators consider this species to be the carbon emitted from oxygen-starved incomplete combustion, commonly from biomass. This makes logical sense because much biomass combustion is not with controlled oxygen supply, especially in the open field burning of postharvest crop residue. The geographical distribution shown in Fig. 4.2 would appear to correlate with biomass combustion as a source. A study by investigators primarily at IIT Kanpur, India, reports that 30% of the light absorption in the Kanpur area was attributed to brown carbon, mostly at a wavelength of 365 nm. They identify high biomass burning in the area as the cause.[10] They also observe particles with an inner core of black carbon, together with an external crust of organic carbon, the crust being classified as brown carbon. Strictly based on the definition of brown carbon, any particle that is brown and UV absorbing ought to be classified as such, with or without an inner core of black carbon. These investigators point out that the crust acts as a "lens" for the core black carbon, presumably through diffraction of the photons. They perform calculations to show that the effect of the crust is additive to the absorption potential of the black carbon core.

Considerable evidence exists to demonstrate a character different from that of black carbon. Whereas black carbon absorbs in the visible and near infrared portions of the spectrum, brown carbon does so in the ultraviolet and the early portion of the visible.

Ultrafine particles (nanomaterials)

Nanomaterials are particles with at least one dimension smaller than 100 nm. For the purposes of calculation of surface area and the like here, particles are assumed to be spheres, even though particulates certainly come in other shapes. As noted earlier, the air quality discipline considers nano to be under 50 nm, and the category under 100 nm is defined as ultrafine. The terms are used here interchangeably, adopting the materials science definition of nanomaterials. However, the air quality terminology of UF will be used for the species. The key property that dominates the behavior of these small particles is the surface-to-volume ratio. For a 100 nm (0.1 μm) diameter sphere, the surface-to-volume ratio is two orders of magnitude greater than for a 10 μm sphere. UF particles can be viewed as a bridge from atomic to bulk scale. Their properties as well are distinctly different. The difference from bulk properties is especially noticeable. All this is primarily due to the dominance of the surface properties over volume. A higher fraction of atoms is on the surface. The surface energy is high and favors most reactions such as catalysis. It also favors the reaction with ambient species such as VOCs. As most nucleating agents emanating from combustion are on the nano scale, the ability to react with airborne species is high.

Imperfections, such as dislocations, are far fewer than in the bulk matrix. Imperfections are the means by which many materials deform and fail under stress. Lack of imperfections makes nanomaterials strong. This is particularly the case for nanocarbon tubes that are now commonly used for strengthening matrices. Similarly, other properties such as optical and magnetic can be dramatically different from bulk properties. These differences are used to advantage in industry. Those aspects will not be delved into here but will be addressed in the chapters on amelioration and policy. This chapter will stick to the issues that pertain to climate and health.

A growing body of evidence points to UF particles as being the most harmful to health. In fact, one study notes that removing the UF component of PM from diesel emissions eliminated the toxic effect on cardiovascular functions.[11, 12] One of the mechanisms is the action of oxygenated organic species such as oxygenated polycyclic aromatic hydrocarbons. They are commonly associated with the combustion of woody biomass and diesel. Interestingly, although the particle itself may be in the fine range, the deposited organic species are often in the UF range. This coating comprising UF particles, often hydrophobic, as opposed to the original substrate being hydrophilic, dominates the oxidative stress responses of $PM_{2.5}$ aerosols, leading to cardiovascular impairment.[13] Ozone is known to play a key role in the oxidative step. The variations of this hybrid character of $PM_{2.5}$ were not actively studied with respect to either cardiovascular or respiratory impacts.[13] Such investigations could be important because the layer, with the UF organics, appears to be the bad actor, and yet the layer could differ in composition and character based on several variables, including nature of chemicals in the atmosphere, intensity of ozone, and photochemical exposure time. Accordingly, the nature of interventions contemplated, be they policy based or engineered solutions, could be affected by that detail.

Bioaerosols

Shifting gears now, consider particulates formed from or by living things. Yes, we are talking about the bioaerosol. The course of human history has been changed by bioaerosols. Far from hyperbole, consider the 1918 influenza pandemic that infected one-third of the world's population and claimed an estimated 20–50 million lives. Or the Irish potato famine caused by the mold *Phytophthora infestans*, which killed over a million people and set a quarter of Ireland's population on an emigration journey that reshaped host countries' cultures. The increasing prevalence of emerging infectious diseases or zoonotic ones such as severe acute respiratory syndrome and avian influenza, respectively, are airborne, contagious pathogens—that is a kind of bioaerosol. Given the obvious and significant potential to affect human health, bioaerosols will be revisited from time to time, so let us set a baseline for our future discourse.

Bioaerosols can be viable, nonviable, or both. Common examples of nonviable bioaerosols are pollen and animal dander; while neither of these is alive, they both can trigger allergies or toxic reactions within an unlucky individual. In contrast, a viable bioaerosol contains biological material capable of multiplying. By this definition, viruses are included as viable bioaerosols (even though there remain divergent viewpoints among microbiologists as to whether viruses are "alive"). Bioaerosols comprise a broad class, but we shall focus on the major categories of microbes: bacteria, viruses, and fungi. Even within these kinds of bioaerosols, there is a diversity of life that defies simple reductionist axioms. The intent is not to provide an exhaustive examination of microbial bioaerosols but merely a flavor of, and an appreciation for, this important category of airborne particulates. Again, a theme consistent with the earlier examination of combustion byproducts: size matters.

Bacteria

These single-celled organisms can be vegetative or, for some species, sporulated. A vegetative bacterium is in a routine, normal state that includes growth by cell division. Vegetative bacteria typically range in size from 0.5 to 10 μm, with most averaging around 2 μm in length. Average is highly subjective, as the varying geometries of different shapes (rods, spheres, and spirals, etc.) produce a range of sizes. Bacteria are small because they need the larger surface area for taking in nutrients from the surrounding environment directly through diffusion. A canonical bacterium, *Escherichia coli*, is rod shaped and about 2 μm long. For comparison, most animal cells are 10–100 μm. A human red blood cell is 8 μm in diameter, and a human egg cell is 100 μm. And, as we all know from personal experience, bacteria are small enough to gain entry into our bodies and reach environments favorable for growth and colonization.

Bacterial spores are a form of Mother Nature's deep hibernation. For some species facing deprivation, usually starvation, a dormant, highly resistant spore forms to protect the cell's genetic material. These endospores, 1–3 μm in diameter, can outlast threats and assaults that would obliterate normal bacteria. Impervious to high heat, ultraviolet irradiation, and many chemicals, endospores are difficult to eliminate. Why does this matter? Because bacteria belonging to the genera *Bacillus* and *Clostridium* are among those that can form spores. The former features prominently in the darkest fears of bioweapons, *Bacillus anthracis*, the causative agent of anthrax; *Clostridium* is the culprit behind a range of diseases from tetanus to botulism to challenging nosocomial infections. Their small size and environmental hardiness mean that spores not only persist in the environment but also can penetrate the deepest reaches of our lungs, where the hospitable conditions reactivate the bacteria. In short, these pose a serious challenge.

Viruses

These microbes only replicate inside living cells. Beyond their intracellular lives, viruses are nothing more than genetic material missiles packaged for a specific cellular target. As with bacteria, there is a staggering viral diversity, but a key common trait is their small size. In fact, the history of how viruses were discovered is linked to the fact that they could not be seen with the microscopes of the day and passed through a filter whose sieve was smaller than bacteria. Viruses are roughly 100 times smaller than bacteria. They live in a nano world, ranging in size from 20 nm to 300 nm. If we take a hypothetical 100 nm virus (an influenza virus ranges from 80 nm to 120 nm), then a single 10 μm particle could contain up to 1 million virions.

This becomes a critical dimension to human health when one considers that the infectious dose (the ID_{50}, a dose whereby half of a population will become infected) for smallpox was 10–100 virions. So, a solitary inhaled particle would be more than enough to result in an infection. Smallpox is not alone in these infection dynamics. Even for a pathogen with an ID_{50} higher by two orders of magnitude, an inhalation of several particles would likely result in infection.

Fungi

The humble fungus is a kingdom—the highest taxonomic classification of living things after domain. Its members range from yeast and molds all the way to mushrooms. Their gifts to humanity include bread, wine, and antibiotics, whereas they wreak devastation with smuts, Dutch elm disease, root rot, and even human infections such as aspergillosis. Fungi often grow in a filamentous form known as hyphae, but it is the spores that most interest bioaerosol aficionados. Fungal spores range in size from 1 μm to 100 μm. As with bacteria, the smaller airborne spores are most readily inhaled and can cause health issues.

Dispersal and penetration

Bioaerosols have two sources, naturally occurring and intentionally released. A cough or sneeze can generate up to 40,000 droplets at speeds up to 200 mph, and the smaller particles can remain aloft and alive for several hours and are ultimately dispersed throughout a room. Wind also generates and spreads bioaerosols, such as spores from the gill of a mushroom and plant pollen. Biodefense experts worry about the deliberate release of bioaerosols as a means of attack. The delivery vehicle could range from the sophisticated bomblet the Soviets constructed to disseminate anthrax spores all the way down to a simple leaf blower or nebulizer. The chilling reality is that the threshold for a bioweapon attack continues to decline. Regardless of whether an exposure is naturally or unnaturally derived, the size plays an important role in how deeply the particles can penetrate the body.

The science of bioaerosols is exceedingly complex and fraught with many dependencies related to the biological material itself and the microenvironment around the particles. Bioaerosols in nature are rarely pure; they are heterogeneous mixtures and often larger than the individual microbe. This size is important because it dictates the mechanism of lung penetration. The total respiratory system deposition for particles from 1 μm to 10 μm is over 80%, with most ending up in the head airways. For the smaller range, 1–3 μm, close to 20% of the total deposition penetrates down into the lung alveoli, where gaseous exchange occurs. Almost 60% of particles of 10–100 μm are deposited in the alveoli. The body has specialized defenses here such as alveolar macrophages because the usual bulwarks of mucus and cilia would prevent gas exchange. As seen with $PM_{2.5}$ in general, these particles can access the bloodstream, thereby opening the highway to the body and broader health impacts. So again, size matters. The importance of being small is underlined for bioaerosol-related morbidity and mortality.

References

1. IEA. *Small increase in energy investment could cut premature deaths from air pollution in half by 2040, says new IEA report.* IEA News; 27 June 2016. Accessed July 22, 2020 https://www.iea.org/news/small-increase-in-energy-investment-could-cut-premature-deaths-from-air-pollution-in-half-by-2040-says-new-iea-report.
2. Stohl A, Klimont Z, Eckhardt S, et al. Black carbon in the Arctic: the underestimated role of gas flaring and residential combustion emissions. *Atmos Chem Phys.* 2013,13(17):8833–8855. https://doi.org/10.5194/acp-13-8833-2013.
3. Lü P, Yu S, Zhu N, et al. Genome encode analyses reveal the basis of convergent evolution of fleshy fruit ripening. *Nat Plants.* 2018;4:784–791.
4. Zhang B, Tieman DM, Jiao C, et al. Chilling-induced tomato flavor loss is associated with altered volatile synthesis and transient changes in DNA methylation. *PNAS.* 2016;113(44):12580–12585. https://doi.org/10.1073/pnas.1613910113.
5. Kim JK, Cho HC, Kim SC. Removal of unburned carbon from coal fly ash using a pneumatic triboelectrostatic separator. *J Environ Sci Health A Tox Hazard Subst Environ Eng.* 2001;36(9):1709–1724. https://doi.org/10.1081/ese-100106253.
6. Yinon L. *Ultrafine Particle Emissions: Comparison of Waste-to-Energy with Coal- and Biomass-Fired Power Plants.* Thesis for M.S. in Earth and Environmental Engineering, Columbia University; January 2010.
7. Lichtveld K, Ebersviller SM, Sexton KG, Vizuete W, Jaspers I, Jeffries HE. In vitro exposures in diesel exhaust atmospheres: resuspension of PM from filters versus direct deposition of PM from air. *Environ Sci Technol.* 2012;46(16):9062–9070. https://doi.org/10.1021/es301431s.
8. Feng Y, Ramanathan V, Kotamarthi VR. Brown carbon: a significant atmospheric absorber of solar radiation? *Atmos Chem Phys.* 2013;13(17):8607–8621. https://doi.org/10.5194/acp-13-8607-2013.
9. Argonne National Laboratory, Environmental Science Division. *Understanding Brown Carbon Aerosols and Their Role in Climate Change*; Accessed July 22, 2020. https://www.evs.anl.gov/research-areas/highlights/brown-carbon.cfm.
10. Shamjad PM, Tripathi SN, Thamban NM, Vreeland H. Refractive index and absorption attribution of highly absorbing brown carbon aerosols from an urban Indian city—Kanpur. *Sci Rep.* 2016;6(1). https://doi.org/10.1038/srep37735.
11. Mills NL, Miller MR, Lucking AJ, et al. Combustion-derived nanoparticulate induces the adverse vascular effects of diesel exhaust inhalation. *Eur Heart J.* 2011;32(21):2660–2671. https://doi.org/10.1093/eurheartj/ehr195.

12. Mills NL, Miller MR. How can we protect susceptible individuals from the adverse cardiovascular effects of air pollution? *Am J Respir Crit Care Med.* 2016;193(9):940–942. https://doi.org/10.1164/rccm.201512-2447ED.
13. Delfino RJ, Staimer N, Tjoa T, et al. Associations of primary and secondary organic aerosols with airway and systemic inflammation in an elderly panel cohort. *Epidemiology.* 2010;21(6). https://doi.org/10.1097/EDE.0b013e3181f20e6c.

Health effects of airborne particulates

Earlier chapters touched on the importance of particulate matter (PM) exposure in health. Many studies have shown negative consequences. This chapter addresses the mechanisms involved in how breathing PM causes stress and sometimes organ failure. Also covered are how changes in size or composition of a particle affect toxicity. The discussion begins with how a single particle affects a human lung cell and then explains how accumulating concentrations of PM lead to morbidities and mortalities globally. The focus is on pediatric health and the changing impact of PM exposure from infancy to adulthood.

Mechanisms of action on organs

A particle's toxicity results from its ability to trigger the formation in cells of reactive oxygen species (ROS) that cause inflammation.[1] Whereas it is clear from epidemiological studies that exposure to PM causes inflammation, the exact mechanisms are still not fully understood. However, the ability of PM to cause damage is known to be a function of size and composition.[2,3] The smaller the particle, the deeper it travels into the lung (Fig. 5.1; particle size in μm).

Larger particles are caught by mucus in the nasal cavity and trachea. In contrast, smaller particles can travel deep into the lungs, into the bronchus, and bronchioles. PM_1 (less than 1 μm) penetrate the epithelial lung cells of the alveoli, causing damage at the cellular level. Huang et al.[4] concluded that the effects of coarser particles were more physiological in nature, whereas those of finer particles were more biological. Cho et al.[5] found that coarser particles caused lung inflammation but no cardiovascular effects, whereas finer particles caused cardiovascular effects but no lung inflammation. These results suggest that the health consequences from PM_1 differ from those of larger particles not only in location but also in nature.

Size appears to directly affect toxicity, as noted in Chapter 4. As diameter decreases, the surface area increases exponentially in proportion. This means

Particulates Matter
https://doi.org/10.1016/B978-0-12-816904-9.00013-1

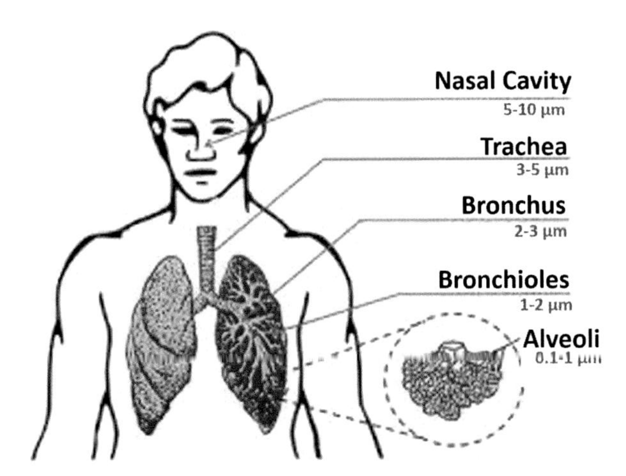

Fig. 5.1 Regional deposition of particles in the respiratory system. *(Modified from Kumar A, Chen F, Mozhi A, et al. Innovative pharmaceutical development based on unique properties of nanoscale delivery formulation.* Nanoscale. *2013;5(18):8307. https://doi.org/10.1039/c3nr01525d (web archive link).)*

that 1 μg of small particles, which can travel deep into the lungs, will have more surface area than 1 μg of large particles. So, they not only travel deeper but also carry larger amounts of toxins on their surfaces that can then be deposited into respiratory organs.[6] A review on deposition rates estimated that approximately 60% of inhaled ultrafine particles ($PM_{0.1}$, less than 0.1 μm, or 100 nm) are deposited in the lung and not exhaled.[3] In vitro, ultrafine particles (1–100 nm) are more toxic than larger particles of similar composition,[7–9] and in vivo, ultrafine particles crossed tissue junctions into cell membranes and damaged cell organelles and induced mutation of DNA.[10–13] The latest research confirmed that ultrafine particles can cross from lung cells into the blood and be transported to virtually any organ system, where the toxins on the particles could cause cardiovascular, lymphatic, neurological, endocrine, and reproductive damage.[2,14]

This creates a concerning scenario because there are no health regulations in place for the smallest particles. Existing regulations are mass based, typically stated in $\mu g/m^3$, and these particles collectively have low mass. The evidence is also not conclusive about the effect of differences in composition on toxicity, although it is broadly known that toxicity can differ from one particle source to another. The science is conclusive that the smallest are the worst offenders, but more causal evidence will be necessary to determine how composition contributes to toxicity and what levels of exposure (if any) could be considered safe.

Morbidity and mortality

In 2016, the World Health Organization (WHO) reported that "91% of the world population is living in places where the WHO air quality guidelines are not met."[15] Coarse particles have been shown to cause lung inflammation that can lead to or worsen asthma, chronic obstructive pulmonary disease (COPD), pneumonia, and lung cancer. Fine particles have been shown to cause cardiovascular damage that can lead to or worsen cardiac arrhythmias, ischemic heart disease, myocardial infarction, vascular dysfunction, hypertension, and atherosclerosis. With so much exposure to unhealthy levels of PM, morbidity and premature mortality are inevitable. And while not all premature deaths from these diseases can be attributed directly to PM exposure, the WHO estimated that ambient PM caused 16% of all lung cancer deaths, 25% of all COPD deaths, 17% of all ischemic heart disease deaths, and 26% of all respiratory infection deaths. Fig. 5.2 shows that mortalities increase linearly or logarithmically as ambient $PM_{2.5}$ increases.[16]

If listed as its own category, PM would be the sixth leading cause of global premature deaths, accounting for 3.8 million annually.[15] A more dire

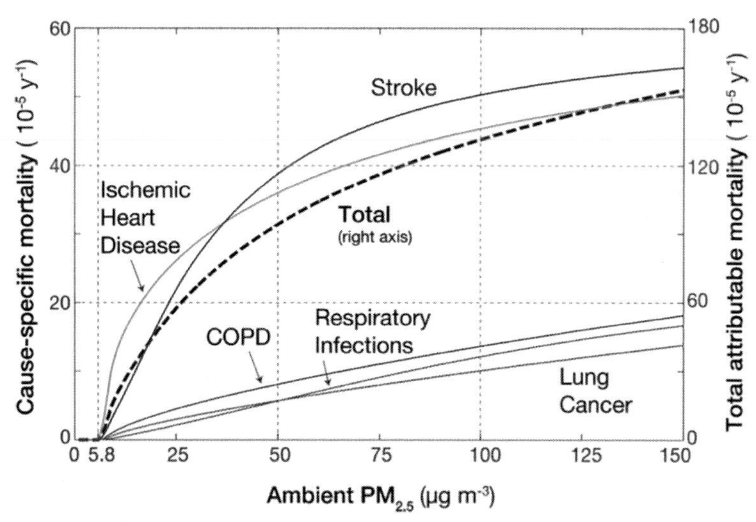

Fig. 5.2 Global concentration-mortality relationships for ambient $PM_{2.5}$ in five diseases.[16] *(Courtesy Apte JS, Marshall JD, Cohen AJ, Brauer M. Addressing global mortality from ambient $PM_{2.5}$.* Environ Sci Technol. *2015;49(13):8057–8066. https://doi.org/10.1021/acs.est.5b01236 (web archive link).)*

conclusion is drawn from new research: this number is likely a major underestimate, and the true number is closer to 8.8 million when PM's role in cardiovascular-related mortality is included.[17,18]

Mortality in the future will depend on the concentration and nature of PM in air. Concentration can be altered by two major influences: anthropogenic emissions and global climate change. In the United States, anthropogenic emissions are predicted to decrease over the next century, leading to a decline in total deaths attributable to PM.[19] However, as the planet will continue to warm even with the reduction in anthropogenic emissions, that decline becomes less significant. Warmer temperatures from climate change are expected to increase frequency and severity of PM-producing wildfires that are projected to increase in frequency so much over the next century that smoke could become the dominant source of $PM_{2.5}$ in the western United States during the summer fire seasons.[20,21] Accordingly, deaths from wildfire-derived PM will increase and could offset the decline in anthropogenic-derived PM deaths by the end of the century.[19]

But there is cause for optimism. An aggressive global $PM_{2.5}$ mitigation program that aligns with the WHO PM standards could avoid an estimated 23% of the 3.8 million annual deaths attributable to PM.[16] Popularization of interventions such as nonwood-burning cookstoves, electric vehicles, and location of electricity production away from populated areas would reduce direct exposure, as discussed in later chapters. Lastly, and counterintuitively, small improvements in cleaner areas could avoid more deaths than small improvements in highly polluted areas, due to the nonlinear concentration-response relationship of PM. Starting small and improving the air quality in cleaner cities could be more effective in reducing the sheer number of premature mortalities.

Long-term systemic impacts

Given the evidence that PM's effects extend beyond the lungs and into the cardiovascular system, new concerns have arisen about other body systems. Because of the ability of ultrafine particles to cross the blood-brain barrier, the brain and central nervous system, the endocrine system, and the reproductive system could be at risk. Research of these questions is new and in some cases inconclusive, such as whether exposure of mothers can cause low birth weight infants. But the evidence for long-term systemic impact is substantial in two chronic diseases, diabetes mellitus and Alzheimer's.

PM provides a surface onto which chemical compounds can attach. One class of compounds that has been found attached is endocrine disruptors that interfere with hormone function and metabolic regulation.[22] Whether the effect is manifested by oxidative stress, endothelial dysfunction, impaired insulin signaling, or increased adiposity has not been confirmed,[23–25] but multiple studies have found an association between high levels of $PM_{2.5}$ and type-2 diabetes.[26,27] The estimated association is that every 10 $\mu g/m^3$ increase in $PM_{2.5}$ in a region causes a 1% increase in diabetes prevalence. To put this in context, the WHO's guideline for maximal annual mean $PM_{2.5}$ is 10 $\mu g/m^3$, and the 50 most polluted cities in the world average over 75 $\mu g/m^3$. The worst, Kanpur, India, is at 173. India has 9 of the 10 most polluted cities in the world.[28]

Alzheimer's disease is a bit trickier to link to PM because the detailed pathogenesis of the disease is still not well understood. In epidemiological studies, living in places with PM levels above the US Environmental Protection Agency's 12 $\mu g/m^3$ $PM_{2.5}$ standard and living within 50 m of a major highway were found to be associated with significantly higher risk of developing dementia.[29,30] Although the association with distance from a highway was demonstrated in the study by Chen et al., ascribing causality to the PM is difficult because of the presence of other pollutants such as NO_2 that were also positively correlated with dementia. In the Framingham Heart Study, a long-term cardiovascular study in New England involving brain MRI scans, investigators at Harvard Medical School found that those living closer to a major roadway had smaller cerebral brain volume.[31] Shortly after that report, Chen et al. at the University of Southern California reported similar brain shrinkage in a study of 1403 elderly females without dementia.[32] They found white matter reduction of 6 cm^3 for every 3.5 $\mu g/m^3$ increase in estimated $PM_{2.5}$ (gray matter appeared unaffected). White matter is the collection of insulated nerve fibers connecting different parts of the brain. This reduction equates to about 1–2 years of normal brain aging.

Disoriented dogs: The first links between PM and dementia

In 2001, a neuroscientist named Lilian Calderón-Garcidueñas living in Mexico City noticed that dogs were becoming disoriented in more polluted parts of the city. The dogs were not recognizing their owners, and when they died, Calderón-Garcidueñas brought their brains to her laboratory to investigate. She found amyloid β-42 peptides in the brains of the dogs—the same peptides that are associated with Alzheimer's disease.[33]

Whereas it is not known whether the peptides are a cause or a consequence of Alzheimer's disease, Calderón-Garciduenas' finding sparked a conversation among scientists about whether it was even possible for air pollution to affect the brain. This was well before the beginning of work on ultrafine and nanoparticles and before the discovery in 2014 that ultrafine particles had the ability to reach the brain and produce harmful ROS. Without a biological mechanism to explain her associations, Calderón-Garciduenas decided to undertake her own studies on the likely underlying biological mechanism. Between 2002 and 2008, she studied children exposed to the ambient air pollution and performed further analyses on the dogs in the city. In both dogs and children, she found chronic inflammation of the upper and lower respiratory tracts, alterations in inflammatory mediators, and disruption of the nasal epithelial barrier and the blood-brain barrier. She also found changes in heart rhythm in the children and pinpointed the early expression of nuclear factor-κB and endothelial- or glial-inducible nitric oxide synthase that can lead to the leakage of inflammatory proteins across the blood-brain barrier.[34]

She continues her work on the pathogenesis by which air pollution contributes to the dementia associated with Alzheimer's and Parkinson's diseases. Recently she has been investigating how neurological damage is caused by air pollution by means of the olfactory bulb in the brain, which is responsible for smell. She has identified the upregulation of cyclooxygenase-2, interleukin-1β, and CD-14 that damage olfactory connections and allow the accumulation of amyloid β-42 and α-synuclein. The accumulation of these proteins leads to abnormalities in the limbic system, specifically a decrease in the number of stem cells in the olfactory bulb. She has also investigated how brain damage can be caused by vagus and trigeminal nerve damage, blood-brain barrier damage, and oxidative damage that result from the central nervous system's response to chronic inflammation.

Two thousand five hundred kilometers to the north, at the University of Southern California, a chemical engineer toils in a trailer near a freeway overpass collecting air mixed with automotive exhaust from about 300,000 vehicles a day. A stream containing just ultrafine PM is separated out and injected into mice genetically engineered to carry the gene for human amyloid β. Amyloid β peptides were observed by Calderón-Garciduenas in the brains of the demented dogs. When compared with the control mice breathing normal air, the test mice showed elevated production of amyloid β and other indications of brain damage correlating with memory loss. Inflammatory molecules associated with Alzheimer's disease were also found.

The promise of the research on the impact of PM on brain function is that if indeed a causal link is found between ultrafine PM and Alzheimer's disease, policymakers will be handed a weapon against a devastating disease that, for the most part, remains untreatable. And we will have the Demented Dogs of Mexico City to thank.

Alongside Calderón-Garcidueñas' publications, many other researchers have added mechanistic findings to the mounting evidence that PM causes significant brain damage. ROS-induced damage to neurons and the effect on neurobehavioral function remain a topic of focus.[35,36] The findings could help explain the associated increased risk of intentional self-harm deaths and nervous system disorders in regions with high $PM_{2.5}$.[37] These studies notwithstanding, investigation of PM impact beyond the respiratory and cardiovascular systems is still nascent. Much remains to be done.

Detection and estimation of episodic insults

PM is in the air almost everywhere, and there appears to be no threshold for having some level of negative cardiovascular and other impacts from exposure.[38] The good news is that associations between exposures and outcomes are such that the reducing the number of particulates can have an immediate significant impact on health. Coughing, wheezing, aggravated asthma, and lung damage can be signs of exposure to an unhealthy amount of PM.[39] Hospital admissions data and death certificates can be used as a proxy for exposure level and to assess whether a community is being subjected to regular exposure at an unhealthy level.

Episodic insults are defined as particulate emissions with a temporal aspect. Of the three major sources of PM emissions, only coal-fired electricity generation is nontemporal because such plants run continuously except for maintenance shutdowns. Wood-burning cookstoves and trash burning are diurnal. Forest fires, although still accounting for a minor fraction of PM, are likely to have a larger presence in a global warming future. Most of these are seasonal. Some nonseasonal occurrences are driven by sparking in aging power transmission lines in the proximity of flammable canopies, such as those of the eucalyptus tree. The third major contributor, automotive exhaust, is temporal for high-intensity periods such as rush hour.

Health effects related to airborne pollutants are of course location specific, such as in "street canyons" formed by tall buildings, where particles are trapped and concentrated.[40] In another location-specific example, studies have shown increased incidence of disease for residences within 200 ft of major roadways. Whereas diesel trucks are considered the main culprits here, gasoline vehicles can be equally or more harmful because of secondary organic aerosols with ROS.[41] Idling, especially of trucks at delivery stops and traffic jams, is an episodic example.

Estimates of the health effects of ozone often consider the periodicity of emissions. In one experiment on lung inflammation, diesel exhaust exposure

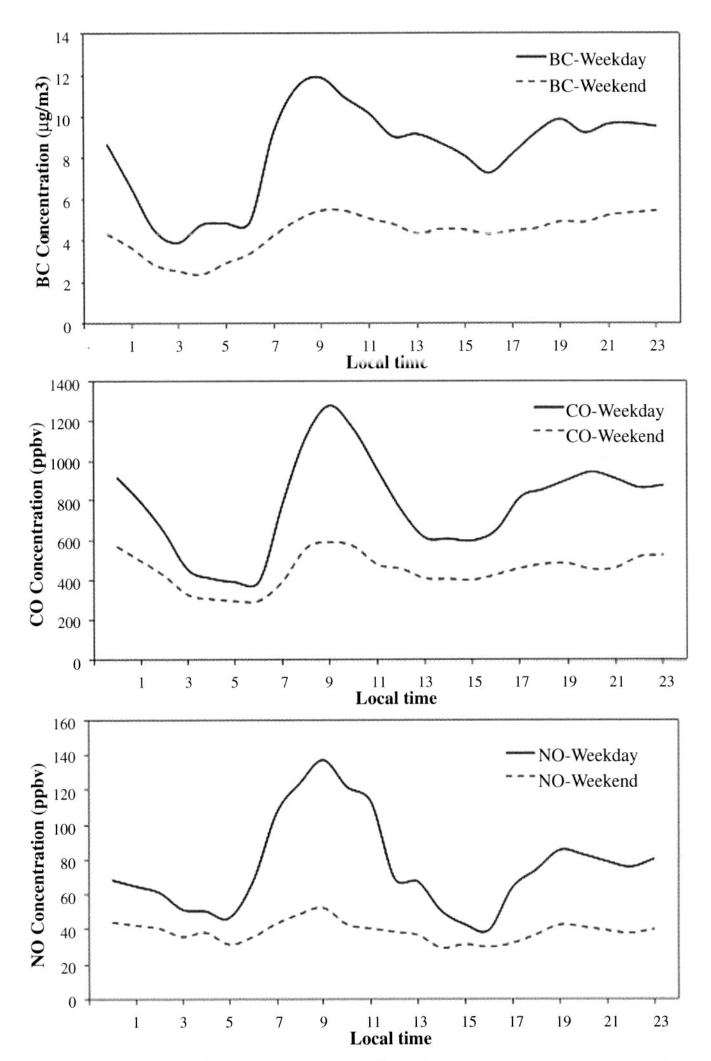

Fig. 5.3 Diurnal variations of concentrations of black carbon, carbon monoxide, and nitric oxide during weekdays and weekends.[43] *(Modified from Behera SN, Balasubramanian R. The air quality influences of vehicular traffic emissions. In: Sallis PJ, ed.* Air Quality Measurement and Modeling*; 2016 [chapter 5]. https://doi.org/10.5772/64692 (web archive link).)*

preceded ozone exposure.[42] This mimicked environmental temporality, in which diesel exhaust peaks during morning rush traffic and ozone peaks in the afternoon. A study in Singapore demonstrated diurnality in black carbon, carbon monoxide, and nitric oxide (Fig. 5.3[43]); the morning rush effect was striking for all three pollutants; the weekend data function as controls.

An interesting case of episodic insult is truck convoys. Oil and gas drilling operations require delivery of materials at regular intervals a few days apart, corresponding to the different phases of operation. The explosion in shale oil and gas drilling has resulted in a steep increase in delivery of water and sand required for hydraulic fracturing. The delivery necessitates long convoys, with many of these traversing small towns that would otherwise experience minimal traffic. These towns receive episodic diesel emissions; virtually all trucks in this category are diesel fueled. In principle, military convoys produce a similar burden, but those episodes are not likely to be frequent. Oil- and gas-related emissions are quite different. A well can be drilled and completed in 2 weeks. Multiple locations can be in the same general vicinity. Accordingly, a town could expect convoys every few days for years. The degree to which this constitutes a health burden merits study. As noted in Chapter 6, portable PM monitors with acceptable accuracy are now available. These could be deployed on the truck routes. But obtaining knowledge of the routes would likely require regulatory intervention. One avenue could be the requirement to notify the state environmental agency of the route details as a precondition for approval of a drilling permit.

Pediatric effects of exposure

Consideration of chronic disease that results from a lifetime of PM exposure tends to focus on cumulative impact upon adults. This is reasonable because adulthood comprises the majority of a lifetime, and so most exposure occurs then. However, an easily overlooked fact is that children experience exposure differently, and because they are still developing, the impact can be more adverse and lasting (Fig. 5.4).

Children are far more susceptible to the immediate impact of PM for several reasons. First, young children have a faster breathing rate than teens and adults, so they inhale more air (and thus more PM) relative to body weight.[44,45] Second, the airways are shorter and narrower that result in inhaled particles having more direct access to lung tissue and obstructing smaller airways.[46] Making matters worse, children spend more time playing outdoors and are more likely to breathe through their mouths, thus inhaling more polluted air that bypasses nasal filtering.[47] The rate of particle deposition on the lung surface area is 35% greater in children than in adults.[48] In addition, the respiratory and immune systems are still developing, so the lungs are more directly affected. Eighty percent of alveoli develop after birth, and $PM_{2.5}$ exposure has been shown to reduce lung growth such that it

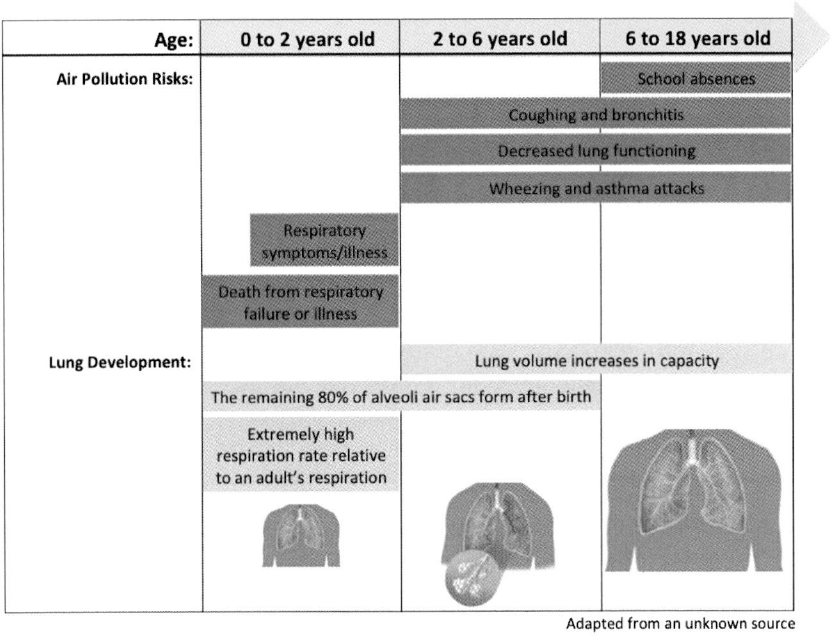

Age: 0 to 2 years old | 2 to 6 years old | 6 to 18 years old
Air Pollution Risks:
School absences
Coughing and bronchitis
Decreased lung functioning
Wheezing and asthma attacks
Respiratory symptoms/illness
Death from respiratory failure or illness
Lung Development:
Lung volume increases in capacity
The remaining 80% of alveoli air sacs form after birth
Extremely high respiration rate relative to an adult's respiration
Adapted from an unknown source

Fig. 5.4 Developmental profile of PM impact on children. *(Modified from anonymous source; see http://fnsb.us/transportation/AQDocs/AQ%20resource%20guide.pdf.)*

never reaches full capacity in adulthood.[47] This impairment of lung-function growth can lead to worse adult lung functioning (as measured by FEV1 and MEF25-75) that can develop into COPD later in life.

Given this high level of susceptibility, it is unsurprising that 8.4% of children in the United States currently have asthma.[49] In communities with the highest background levels of PM, the rate of young adults with clinically significant reduced lung function ($<$80% predicted) is fivefold higher than in other communities.[50] The epidemiological evidence for the association between high PM levels and increased asthmatic episodes was confirmed by numerous studies, and research is underway to isolate the role of different sized PMs on asthma. In 2008, Tecer et al. showed that all size fractions were significantly associated with asthma, although coarse PM was a better predictor of hospital admissions than fine PM.[51] This finding was confirmed by McCormack et al., who considered how exposure to indoor coarse PM was correlated with the use of rescue medications for asthmatic preschoolers.[52] For outdoor coarse PM, Keet found a similar trend in that each 1 $\mu g/m^3$ increase was associated with a 0.6% increase in asthma prevalence, a 2.3% increase in hospitalizations, and a 1.7% increase in emergency room visits.[53]

In addition to asthma, PM exposure places children at risk for a plethora of other adverse consequences. Those exposed to high PM-emitting fuels are more likely to develop bacterial infections such as pneumonia. To exacerbate matters, a bacterial respiratory infection also makes children more susceptible to the effects of air pollution.[54–56] This is believed to be due to PM's downregulation of airway antimicrobial proteins and peptides such as salivary agglutinin and increased IL-8 and TNF-α that accelerate pathogenic infections.[57] Children near the tropics where ambient pollution is high develop vitamin D-deficiency rickets because the hazy air prevents ultraviolet-B rays from reaching the ground level to assist with the synthesis of vitamin D3.[58] Also, interestingly, a recent study found that healthy vitamin D levels can protect children against asthma in high-PM environments through antioxidant and immune-modulatory pathways.[59] So, these children were both vitamin D deficient because of pollution and more susceptible to the effects of pollution because of their deficiency! Recent research has found associations in children between $PM_{2.5}$ and systolic blood pressure, $PM_{2.5}$ and autism risk or severity, PM_{10} and hypertension, and $PM_{2.5}$ and sudden infant death syndrome.[60–63] Work is being done to uncover the biological mechanisms by which PM induces these effects and how to best protect children from them, despite continuing high levels of exposure worldwide.

References

1. Val S, Liousse C, Doumbia EHT, et al. Physico-chemical characterization of African urban aerosols (Bamako in Mali and Dakar in Senegal) and their toxic effects in human bronchial epithelial cells: description of a worrying situation. *Part Fibre Toxicol.* 2013;10 (1):10. https://doi.org/10.1186/1743-8977-10-10.
2. Kang D, Kim J-E. Fine, ultrafine, and yellow dust: emerging health problems in Korea. *J Korean Med Sci.* 2014;29(5):621–622. https://doi.org/10.3346/jkms.2014.29.5.621.
3. Lighty JAS, Veranth JM, Sarofim AF. Combustion aerosols: factors governing their size and composition and implications to human health. *J Air Waste Manage Assoc.* 2000;50 (9):1565–1618. https://doi.org/10.1080/10473289.2000.10464197.
4. Huang S-L, Hsu M-K, Chan C-C. Effects of submicrometer particle compositions on cytokine production and lipid peroxidation of human bronchial epithelial cells. *Environ Health Perspect.* 2003;111(4):478–482.
5. Cho S-H, Tong H, McGee JK, Baldauf RW, Todd Krantz Q, Ian Gilmour M. Comparative toxicity of size-fractionated airborne PM collected at different distances from an urban highway. *Environ Health Perspect.* 2009;117(11):1682–1689. https://doi.org/10.1289/ehp.0900730.
6. Ahn J, Lee JS, Yang KM. Ultrafine particles of *Ulmus davidiana var. japonica* induce apoptosis of gastric cancer cells via activation of caspase and endoplasmic reticulum stress. *Arch Pharm Res.* 2014;37(6):783–792. https://doi.org/10.1007/s12272-013-0312-2.
7. Carlson C, Hussain SM, Schrand AM, et al. Unique cellular interaction of silver nanoparticles: size-dependent generation of reactive oxygen species. *J Phys Chem B.* 2008;112(43):13608–13619. https://doi.org/10.1021/jp712087m.

8. Napierska D, Thomassen LCJ, Rabolli V, et al. Size-dependent cytotoxicity of monodisperse silica nanoparticles in human endothelial cells. *Small.* 2009;5(7):846–853. https://doi.org/10.1002/smll.200800461.
9. Pan Y, Neuss S, Leifert A, et al. Size-dependent cytotoxicity of gold nanoparticles. *Small.* 2007;3(11):1941–1949. https://doi.org/10.1002/smll.200700378.
10. Donaldson K, Brown D, Clouter A, et al. The pulmonary toxicology of ultrafine particles. *J Aerosol Med.* 2002;15(2):213–220. https://doi.org/10.1089/089426802320282338.
11. Hoshino Y, Koide H, Urakami T, et al. Recognition, neutralization, and clearance of target peptides in the bloodstream of living mice by molecularly imprinted polymer nanoparticles: a plastic antibody. *J Am Chem Soc.* 2010;132(19):6644. https://doi.org/10.1021/ja102148f. https://webcms.pima.gov/UserFiles/Servers/Server_6/File/Government/Environmental%20Quality/Air/Air%20Monitoring/AWhatisParticulateMatter1.pdf.
12. Salnikov V, Lukyánenko YO, Frederick CA, Lederer WJ, Lukyánenko V. Probing the outer mitochondrial membrane in cardiac mitochondria with nanoparticles. *Biophys J.* 2007;92(3):1058–1071. https://doi.org/10.1529/biophysj.106.094318.
13. Wilson OM, Knecht MR, Garcia-Martinez JC, Crooks RM. Effect of Pd nanoparticle size on the catalytic hydrogenation of allyl alcohol. *J Am Chem Soc.* 2006;128(14):4510–4511. https://doi.org/10.1021/ja058217m.
14. Akhtar US, Rastogi N, McWhinney RD, et al. The combined effects of physicochemical properties of size-fractionated ambient PM on in vitro toxicity in human A549 lung epithelial cells. *Toxicol Rep.* 2014;1:145–156. https://doi.org/10.1016/j.toxrep.2014.05.002.
15. World Health Organization. *Air Pollution Levels Rising in Many of the World's Poorest Cities.* World Health Organization; 2016. Retrieved from: http://www.who.int/news-room/detail/12-05-2016-air-pollution-levels-rising-in-many-of-the-world-s-poorest-cities.
16. Apte JS, Marshall JD, Cohen AJ, Brauer M. Addressing global mortality from ambient $PM_{2.5}$. *Environ Sci Technol.* 2015;49(13):8057–8066. https://doi.org/10.1021/acs.est.5b01236.
17. Hoek G, Krishnan RM, Beelen R, et al. Long-term air pollution exposure and cardio-respiratory mortality: a review. *Environ Health.* 2013;12:43. https://doi.org/10.1186/1476-069X-12-43.
18. Lelieveld J, Klingmüller K, Pozzer A, et al. Cardiovascular disease burden from ambient air pollution in Europe reassessed using novel hazard ratio functions. *Eur Heart J.* 2019;40(20):1590–1596. https://doi.org/10.1093/eurheartj/ehz135.
19. Ford B, Val Martin M, Zelasky SE, et al. Future fire impacts on smoke concentrations, visibility, and health in the contiguous United States. *GeoHealth.* 2018. https://doi.org/10.1029/2018GH000144. Accessed 11 July 2018.
20. Liu JC, Mickley LJ, Sulprizio MP, et al. Particulate air pollution from wildfires in the western US under climate change. *Clim Change.* 2016;138(3–4):655–666. https://doi.org/10.1007/s10584-016-1762-6.
21. Yue X, Mickley LJ, Logan JA, Kaplan JO. Ensemble projections of wildfire activity and carbonaceous aerosol concentrations over the western united states in the mid-21st century. *Atmos Environ.* 2013;77:767–780. https://doi.org/10.1016/j.atmosenv.2013.06.003.
22. Shaikh S, Jagai JS, Ashley C, Zhou S, Sargis RM. Underutilized and under threat: environmental policy as a tool to address diabetes risk. *Curr Diab Rep.* 2018;18(5):25. https://doi.org/10.1007/s11892-018-0993-5.
23. Baron AD, Steinberg HO, Chaker H, Leaming R, Johnson A, Brechtel G. Insulin-mediated skeletal muscle vasodilation contributes to both insulin sensitivity and

responsiveness in lean humans. *J Clin Invest.* 1995;96(2):786–792. https://doi.org/10.1172/JCI118124.
24. Liu C, Ying Z, Harkema J, Sun Q, Rajagopalan S. Epidemiological and experimental links between air pollution and type 2 diabetes. *Toxicol Pathol.* 2012;41(2):361–373. https://doi.org/10.1177/0192623312464531.
25. Sun Q, Wang A, Jin X, et al. Long-term air pollution exposure and acceleration of atherosclerosis and vascular inflammation in an animal model. *JAMA.* 2005;294(23): 3003–3010. https://doi.org/10.1001/jama.294.23.3003.
26. Heindel JJ, Blumberg B, Cave M, et al. Metabolism disrupting chemicals and metabolic disorders. *Reprod Toxicol.* 2017;68:3–33. https://doi.org/10.1016/j.reprotox.2016.10.001.
27. Weinmayr G, Hennig F, Fuks K, et al. Long-term exposure to fine PM and incidence of type 2 diabetes mellitus in a cohort study: effects of total and traffic-specific air pollution. *Environ Health.* 2015;14:53. https://doi.org/10.1186/s12940-015-0031-x.
28. World Health Organization. *WHO Global Ambient Air Quality Database (Update 2018)*; 2018.
29. Cacciottolo M, Wang X, Driscoll I, et al. Particulate air pollutants, APOE alleles and their contributions to cognitive impairment in older women and to amyloidogenesis in experimental models. *Transl Psychiatry.* 2017;7(1). https://doi.org/10.1038/tp.2016.280.
30. Chen H, Kwong JC, Copes R, et al. Living near major roads and the incidence of dementia, Parkinson's disease, and multiple sclerosis: a population-based cohort study. *Lancet.* 2017;389(10070):718–726. https://doi.org/10.1016/S0140-6736(16)32399-6.
31. Chung M, Wang DD, Rizzo AM, et al. Association of PNC, BC, and $PM_{2.5}$ measured at a central monitoring site with blood pressure in a predominantly near highway population. *Int J Environ Res Public Health.* 2015;12(3):2765–2780.
32. Chen JC, Wang X, Wellenius GA, et al. Ambient air pollution and neurotoxicity on brain structure: evidence from women's health initiative memory study. *Ann Neurol.* 2015;78(3):466–476. https://doi.org/10.1002/ana.24460. Epub 2015 Jul 28 26075655.
33. Calderón-Garcidueñas L, Maronpot RR, Torres-Jardon R, et al. DNA damage in nasal and brain tissues of canines exposed to air pollutants is associated with evidence of chronic brain inflammation and neurodegeneration. *Toxicol Pathol.* 2003;31(5):524–538. https://doi.org/10.1080/01926230390226645.
34. Block ML, Calderón-Garcidueñas L. Air pollution: mechanisms of neuroinflammation and CNS disease. *Trends Neurosci.* 2009;32(9):506–516. https://doi.org/10.1016/j.tins.2009.05.009. Epub 2009 Aug 26. Review 19716187.
35. Gillespie P, Tajuba J, Lippmann M, Chen L-C, Veronesi B. PM neurotoxicity in culture is size-dependent. *Neurotoxicology.* 2013;36:112–117. https://doi.org/10.1016/j.neuro.2011.10.006.
36. Xu X, Ha SU, Basnet R. A review of epidemiological research on adverse neurological effects of exposure to ambient air pollution. *Front Public Health.* 2016;4. https://doi.org/10.3389/fpubh.2016.00157.
37. Li T, Yan M, Sun Q, Brooke Anderson G. Mortality risks from a spectrum of causes associated with wide-ranging exposure to fine PM: a case-crossover study in Beijing, China. *Environ Int.* 2018;111(February):52–59. https://doi.org/10.1016/j.envint.2017.10.023.
38. Ware JH. Particulate air pollution and mortality—clearing the air. *New England Journal of Medicine.* 2000;343(24):1798–1799. https://doi.org/10.1056/NEJM200012143432409.
39. Pima.gov. "What is PM?" pdf.
40. Kumar P, Robins A, Vardoulakis S, Britter R. A review of the characteristics of nanoparticles in the urban atmosphere and the prospects for developing regulatory controls. *Atmos Environ.* 2010;44:5035–5052.

41. Platt SM, El Haddad I, Pieber SM, et al. Gasoline cars produce more carbonaceous particulate matter than modern filter-equipped diesel cars. *Sci Rep*. 2017;7(1). https://doi.org/10.1038/s41598-017-03714-9.
42. Bosson J, Pourazar J, Forsberg B, Ädelroth E, Sandström T, Blomberg A. Ozone enhances the airway inflammation initiated by diesel exhaust. *Respir Med*. 2007;101 (6):1140–1146. https://doi.org/10.1016/j.rmed.2006.11.010.
43. Behera SN, Balasubramanian R. The air quality influences of vehicular traffic emissions. In: Sallis PJ, ed. *Air Quality Measurement and Modeling*; 2016. https://doi.org/10.5772/64692 [chapter 5].
44. Children are Not Little Adults; 2017. Available from http://www.who.int/ceh/capacity/Children_are_not_little_adults.pdf.
45. Dunea D, Iordache S, Pohoata A. Fine particulate matter in urban environments: a trigger of respiratory symptoms in sensitive children. *Int J Environ Res Public Health*. 2016;13:1246.
46. Gauderman WJ, Urman R, Avol E, et al. Association of improved air quality with lung development in children. *N Engl J Med*. 2015;372:905–913.
47. American Lung Association. *Children and air pollution*; 2003. Available from American Lung Association website https://www.lung.org/our-initiatives/healthy-air/outdoor/air-pollution/children-and-air-pollution.html. Retrieved 27 August 2019.
48. Bennett WD, Zeman KL. Desposition of fine particles in children spontaneously breathing at rest. *Inhal Toxicol*. 1998;10:831–842.
49. Centers for Disease Control and Prevention. *FastStats*; 2017. Available from https://www.cdc.gov/nchs/fastats/asthma.htm. Retrieved 28 August 2019.
50. Grigg J. Particulate matter exposure in children. *Proc Am Thorac Soc*. 2009;6(7):564–569 https://doi.org/10.1513/pats.200905-026RM.
51. Tecer LH, Alagha O, Karaca F, Tuncel G, Eldes N. Particulate matter (PM2.5, PM10-2.5, and PM10) and children's hospital admissions for asthma and respiratory diseases: a bidirectional case-crossover study. *J Toxicol Environ Health A*. 2008;71(8):512–520. https://doi.org/10.1080/15287390801907459.
52. McCormack MC, Breysse PN, Matsui EC, et al. Indoor particulate matter increases asthma morbidity in children with non-atopic and atopic asthma. *Ann Allergy Asthma Immunol*. 2011;106(4):308–315. https://doi.org/10.1016/j.anai.2011.01.015.
53. Keet CA, Keller JP, Peng RD. Long-term coarse particulate matter exposure is associated with asthma among children in Medicaid. *Am J Respir Crit Care Med*. 2017;197 (6):737–746. https://doi.org/10.1164/rccm.201706-1267OC.
54. Barnett AG, Williams GM, Schwartz J, et al. Air pollution and child respiratory health: a casecrossover study in Australia and New Zealand. *Am J Respir Crit Care Med*. 2005;171:1272–1278.
55. Dherani M, Pope D, Mascarenhas M, Smith KR, Weber M, Bruce N. Indoor air pollution from unprocessed solid fuel use and pneumonia risk in children aged under five years: a systematic review and metaanalysis. *Bull World Health Organ*. 2008;86:390–398.
56. World Health Organization. *The Effects of Air Pollution on Children's Health and Development: A Review of the Evidence*; 2005. E86575. Available at http://www.euro.who.int/document/E86575.pdf.
57. Zhang S, Huo X, Zhang Y, Huang Y, Zheng X, Xu X. Ambient fine particulate matter inhibits innate airway antimicrobial activity in preschool children in e-waste areas. *Environ Int*. 2019;123:535–542. https://doi.org/10.1016/j.envint.2018.12.061.
58. Agarwal KS, Mughal MZ, Upadhyay P, Berry JL, Mawer EB, Puliyel JM. The impact of atmospheric pollution on vitamin D status of infants and toddlers in Delhi, India. *Arch Dis Child*. 2002;87(2):111–113. https://doi.org/10.1136/adc.87.2.111.
59. Bose S, Diette GB, Woo H, et al. Vitamin D status modifies the response to indoor particulate matter in obese urban children with asthma. *J Allergy Clin Immunol Pract*. 2019;7 (6):1815–1822.e2. https://doi.org/10.1016/j.jaip.2019.01.051.

60. Buka I, Koranteng S, Osornio-Vargas AR. The effects of air pollution on the health of children. *Paediatr Child Health*. 2006;11(8):513–516.
61. Geng R, Fang S, Li G. The association between particulate matter 2.5 exposure and children with autism spectrum disorder. *Int J Dev Neurosci*. 2019;75:59–63. https://doi.org/10.1016/j.ijdevneu.2019.05.003.
62. Heft-Neal S, Burney J, Bendavid E, Burke M. Robust relationship between air quality and infant mortality in Africa. *Nature*. 2018. https://doi.org/10.1038/s41586-018-0263-3.
63. Zhang Z, Dong B, Li S, et al. Exposure to ambient particulate matter air pollution, blood pressure and hypertension in children and adolescents: a national cross-sectional study in China. *Environ Int*. 2019;128:103–108. https://doi.org/10.1016/j.envint.2019.04.036.

PART THREE

Detecting PM and in vitro techniques to measure health impact

Part Two described how PM forms and underlined the importance of particle size in the impact on health. Having established causality between particles in the air and disease, the focus shifts in this section to measuring the concentrations in both the ambient and occupational settings. Measurement over time and space is required to acquire data to inform mitigation. Discrete measurements at stations are the primary means. Because this can be prohibitive in cost, satellites are used to link to and augment terrestrial measurements. Personal monitors are key for quantifying occupational exposure.

In vitro methods for estimating toxicity of airborne PM rely on collecting the particles on filters and resuspending in a fluid medium for reacting with tissue. Resuspension tends to agglomerate ultrafine particles, and this blurs the impact of mechanisms driven by very small size. An air-liquid interface technique is described, which more realistically mimics the action in lungs and overcomes this shortcoming of the standard method.

Detection and evaluation of airborne particulates

Detection and evaluation of airborne particulate matter (PM) falls into two major categories: spatial distribution over large areas, and local collection, identification, and estimation of quantities. The latter will be dealt with first.

Measurement comprises estimation of concentration by mass and by size distribution (Fig. 6.1). Concentration by numbers of particles is a relatively recent addition to the measurement spectrum and is not yet present in regulations except to a degree in Europe. In some ways, this separation is artificial. Concentration is usually measured in fractions separated by mass, for which size is a reasonable proxy when the species are known. But a separate discussion is merited because the measurement techniques are distinctly different except in certain elements, such as impactors, where they overlap.

Gravimetric estimation

In the simplest form, air is drawn through a filter sized to collect particles of a certain size distribution. The two most common filters are for $PM_{2.5}$ and PM_{10} (particles with diameter $\leq 2.5\ \mu m$ and $\leq 10\ \mu m$, respectively). The filter is weighed before and after collection to estimate the concentration in air. When toxicity measurements are desired, the particles are extracted into a fluid medium. This is the standard means for estimates of biological impact and is known as resuspension. Gravimetric estimation is the standard against which other methods are judged.

A variant of the gravimetric method is manipulation of the input to produce an initial separation of sizes. The most common means for this is the impactor. Cascade impactors were investigated for decades prior to the forerunner of today's devices, which was invented during World War II in England by K.R. May. It was designed to separate out constituents of chemical warfare agents.[1] Essentially, the same principle is used now to separate aerosols by mass. Fig. 6.2 shows such a device.

The particulate-laden air is sucked into the impactor by a vacuum pump, aspirator, or other suction device. Each stage comprises a plate upon which

Particulates Matter
https://doi.org/10.1016/B978-0-12-816904-9.00010-6

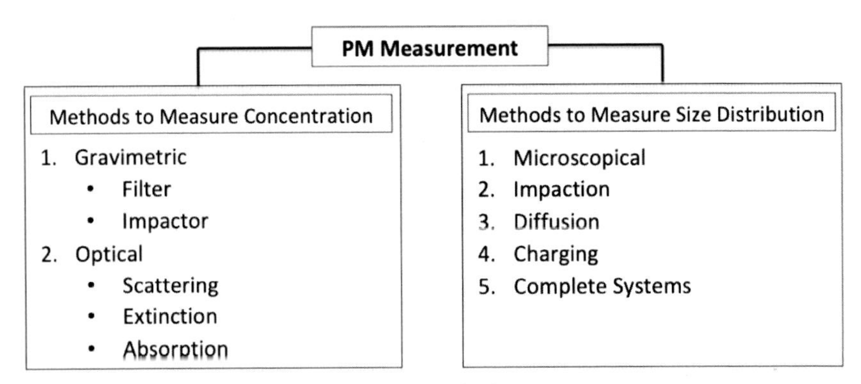

Fig. 6.1 Particulate matter measurement methods.

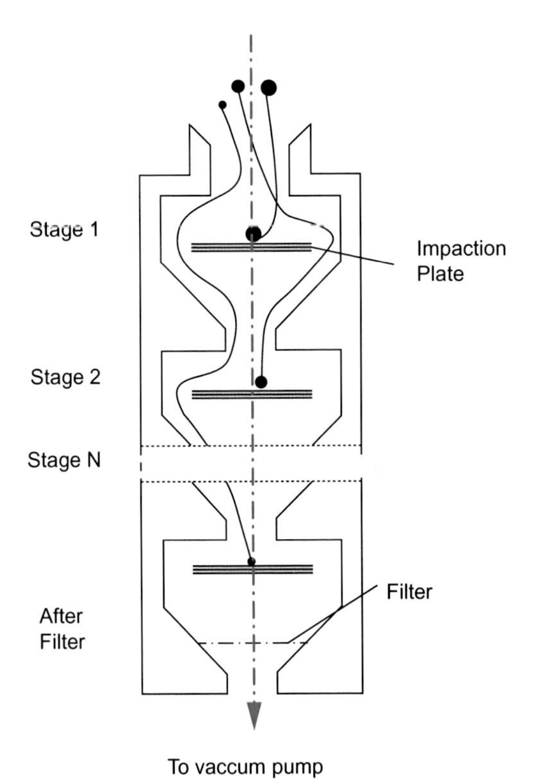

Fig. 6.2 A cascade impactor.

the particle stream impacts after exiting a flow restriction. In modern versions, the flow restriction may comprise multiple nozzles. Larger particles collect on the plate, and smaller ones are diverted down to the next stage, where another mass fraction is collected, and so on. Successive flow restrictions are smaller, causing the increased velocity needed to collect the ever-smaller particles onto the plates. At the end, a filter collects all remaining particles.

The plates in May's device were of glass. Today, they may be made from plastics such as polytetrafluoroethylene (trade name Teflon), polycarbonate, or polyethylene terephthalate (a common plastic for bottles). Sometimes other materials such as aluminum are used. The choice depends on the environment and the final intended investigation of the sample. Common impactors have a last cutoff at 0.3 μm. As discussed in earlier chapters, an increasing focus on ultrafine particles produces a need for impactors to select well into the sub 0.1 μm range.

The micro-orifice uniform deposit impactor exemplifies the sophistication now available. Each substrate, rather than being stationary as shown in Fig. 6.2, can be rotated by a stepping motor. Rotation allows more uniform deposition, and the overall design of the multistage impactor provides a high degree of separation of the sizes. A typical distribution of cuts is 0.010, 0.018, 0.032, 0.056, 0.10, 0.18, 0.32, 0.56, 1.0, 1.8, 3.2, 5.6, 10, and 18 μm. Stages may be removed if the interest is only in a certain number and size of cuts.

The relatively recent availability of impactors offering sub 0.1 μm cuts is timely, considering the evidence presented in Chapter 5 for the disproportionate physiologic impact of the ultrafine component of $PM_{2.5}$, and the cut is increasingly believed to be the principal cause of mortality. The new impactor capability is even more useful for air–liquid interface devices, as will be discussed later in this chapter, than it is for gravimetric methods followed by resuspension in liquid.

Optical methods

Optical methods of estimating particle count and mass are premised upon the action of light impinging on the particle. Some of the intensity may be absorbed and some reflected. The degree to which this happens depends in part on the wavelength of the incident light when compared with the size of the particle. This relationship relies on the size parameter α, defined as

$$\alpha = \frac{\pi D_P}{\lambda}, \tag{6.1}$$

where D_p is the diameter of a spherical particle and λ is the wavelength of incident light. While particles are often misshapen, sphericity is an approximation used in modeling and in the classifications of PM.

Scattering may be classified into three distinct regions:

$\alpha \ll \lambda$ Raleigh scattering,

$\alpha \sim \lambda$ Mie scattering,

$\alpha \gg \lambda$ Simplified geometric scattering.

Only Mie scattering will be discussed here because most optical particle counters operate in roughly that regime due to visible and near-infrared being the workhorse wavelengths used. The relationship between particle size and scattered intensity is generally derived empirically using particles of known sizes. This is because the relationship is nonlinear, in part because real-world particles are off-round. Portable units commonly use optical scattering. The light beam is a substantially coherent output of a laser diode. Portable counters operate in specific PM size bands because of the limitations placed by the dynamic range of the associated detector system.

Counters based on nephelometry

Nephelometry is the principle of estimating PM suspended in a medium. When a light beam is incident on the particles, it is scattered and strikes photomultiplier devices placed at some angle to the beam path. This angle is usually 90 degrees but may also be 70 degrees or another angle at which the maximum scattering intensity is observed. Each particle produces a scattered portion of the incident beam, and the collective intensity is some combination of number of particles and their mass. The same principle is used to measure turbidity of a fluid. In that case, the measured intensity is roughly on the same axis as the incident beam.

Small portable devices are needed to measure PM in many real-life situations. This is particularly the case in studying the occupational health of workers at risk in PM-producing environments such as diesel engine repair shops. When improvements were made to wood-burning cookstoves, the air in the vicinity of cooks was measured to estimate the degree of improvement.[2] Most small, inexpensive devices are based on optical measurements. The simplest ones such as the Clarity P1 are particle counters. The mass is deduced from the particle count. These cost less than USD 1000. More accurate and expensive ones (less than USD 3000) are based on the principle of nephelometry. All are effective only at low absolute loading in the fluid medium. A typically observed ambient loading of 10–200 μg/m^3 is suited to these devices. As a frame of reference, the current World Health Organization guideline for $PM_{2.5}$ is an upper limit of 10 μg/m^3.

Low-cost portable monitors have held the promise of high granularity in measurement and the possibility of a volunteer citizen force to collect data.[3–5] In recognition of the need for such devices, the US Environmental Protection Agency (EPA) conducted tests for the most available ones costing less than USD 2500 (United States Environmental Protection Agency 2016). The comparisons were made against a Grimm Technologies model EDM180 $PM_{2.5}$ (EQPM-0311-195) dust monitor and an R. M. Young model 41382VC relative humidity and temperature sensor. These together comprised the federal equivalent method (FEM) standard. Linear regression analyses were run against the standard under varying ambient conditions. The coefficient of determination, R^2, was the key parameter (R^2 closer to 1 is desired). Table 6.1 gives the results. Only three met the threshold of R^2 values. Of these, the Dylos DC1100 was just a particle counter. The other two units, both nephelometry based, were the Met One 831 and the MicroPEM from RTI International.

The Met One 831 performs better but is more expensive than the MicroPEM. A recent study compared the MicroPEM with a more expensive nephelometry-based device, the MIE pDR-1500 (USD 5000), and a particle counter, the Clarity P1.[6] The Clarity P1 was a USD 100 prototype; more recent information has the commercial version priced at USD 700. The other two were used in multiple studies in the field. As expected, the particle counter undercounted the other two. The takeaway is that portable devices have high performance variability based upon ambient conditions, so the one most fit for the purpose needs to be used. The MicroPEM stands out as the cost-effective alternative when precision is required. The correlation with the FEM is excellent in the laboratory, with R^2 approaching 0.99. But the intended application of these devices is in the field, and so the 0.72 value from the EPA study is more representative. Interdevice variability is low, but frequent recalibration is required.

RTI MicroPEM version 3.2

This is the model that was tested by Fisher et al. It has a two-stage impactor at the inlet, which can be sized for PM_{10}, PM_4, or $PM_{2.5}$. Air flow is initiated by a pump at the rate of 0.5 L/min. The particles may be collected

Table 6.1 Results of US EPA evaluation of portable optical particle counters costing less than USD 2500.

Sensor	R^2	Response	RH limit	Temperature effect	Time resolution	Uptime	Ease of installation	Ease of operation	Mobility
AirBase CanarIT (μg/m³)	0.004	−0.101	100%	None	20 s	Excellent	Good	Excellent	Very good
CairClip PM (μg/m³)	0.064	−0.229	95%	0.657	1 min	Excellent	Good	Very good	Excellent
Carnegie Mellon Speck (particle counts)	0	0.06	90%	None	1 s	Very good	Good	Fair	Good
Dylos DC1100 (particle counts)	0.548	21,368	95%	None	1 min	Very good	Good	Good	Poor
Met One 831 (μg/m³)	0.773	0.049	90%	None	1 min	Excellent	Good	Good	Good
RTI MicroPEM (μg/m³)	0.720	1.35 ± 0.12	95%	0.588	10 s	Very good	Good	Fair	Fair
Sensaris Eco PM (μg/m³)	0.315	0.034	100%	0.313	Unknown	Bad	Poor	Bad	Poor
Shinyei PMS-SYS-1 (μg/m³)	0.152	0.292	95%	None	1 s	Good	Fair	Good	Fair

Courtesy of US Environmental Protection Agency.

for future gravimetric or other analyses. Both battery and online power can be used. The investigation range extends to 5000 μg/m^3, but one might want any nephelometric device to operate with better precision at the lower limit of that range. This is not much of a limitation because the ranges of interest are 1–200 μg/m^3. The impactor limit of $PM_{2.5}$ is likely driven by the current regulation being in that range. The current EPA limit is 12 μg/m^3, revised down from 15 μg/m^3 in 2012. Relaxation of that figure is expected during the current US administration, despite recent studies indicating high morbidity and mortality between 2.8 and 13.2 μg/m^3.[7] Impactors need to be substituted to study the ultrafine range.

The model has the interesting feature of an onboard accelerometer. It documents compliance with the protocol when worn on the person of experimental subjects.[2] The measured data are transmitted through bluetooth to a storage medium. The system is operated using a 780-nm laser diode, and the detection is with an Opti diode photometer with a built-in preamp. Such details are usually a trade-off among performance, size, and cost and can be expected to change with user's requirements without significant technical hurdles. Fig. 6.3 shows the device worn on a camera sling.

Fig. 6.3 MicroPEM Version 3.2 worn on a camera sling.

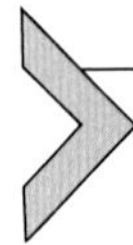

Measuring black carbon

Aethalometers

Because many combustion processes also produce organics, black carbon does not often remain in that state, transforming to brown carbon through the deposition of organic layers. However, it is more likely in particulates from coal combustion. In the case of automotive engines, any estimate of black carbon would likely need to be close to the combustion chamber. Black carbon requires study because its climate change effects are different from those of other PM. It is a strong absorber of light over a broad range of wavelengths. This property is used in estimating black carbon concentrations.

One of the most common means is the aethalometer. It was invented and the word was coined from the Greek meaning "to blacken with soot" by Hansen and colleagues at Lawrence Berkeley Laboratories.[8] The description is apt because the principle of the device is to capture airborne particulates on a filter and measure the transmissivity of light through the filter. In one embodiment, the filter is on a roll and a segment is in the gas stream until a certain loading is reached, upon which the next segment is rolled into place. Incident light will either scatter or absorb, the rest passing through. Nephelometry, discussed above, uses the property of light scatter. Aethalometry is the other portion transmitting through, attenuation being the measurement of interest.

Attenuation through the filter is inversely proportional to wavelength. Modern aethalometers operate between 370 nm (near ultraviolet) and 970 nm (near infrared). Black carbon will absorb at the same rate at all wavelengths. However, if some of the particles tend toward brown carbon due to organic deposits (see Chapter 4), then increased absorption will occur at shorter wavelengths.

Photoacoustic methods

The technique is premised on sound waves being generated upon incidence of high-intensity radiation. The light is absorbed and creates heat. Thermal relaxation of the material generates an acoustic wave. The amplitude of the wave is directly proportional to the thermal mass of the particle, when the particle is small enough. The device is usually in a tube configuration, and the amplitude is amplified nearly two orders of magnitude using resonance techniques, not dissimilar to the effect of organ pipes in a church. This amplification allows the use of conventional microphone pickups.

This in-line technique overcomes a shortcoming of the aethalometer, wherein sample deposition on the filter could alter the optical properties from those of an isolated particle. Whereas this discussion is centered on black carbon detection, the technique is broadly applicable to detecting and quantifying other species.

Size measurement

Because size is known to matter, measurement of the parameter is important. This section will not comprehensively describe all possible options, such detail being available in review articles such as that by Amaral et al.[9] The diameter is the most commonly approximated parameter, despite the particles themselves seldom being round. The physical properties used are geometry, electrical mobility, inertia, and response to incident light.

Many of the measurement means are described earlier in this chapter, where the focus is on particle concentration, but the techniques also elicit mass. For example, methods using incident light rely in part on the property that the interaction is more efficient when the wavelength is close to the particle size.

Also already discussed is the use of cascade impactors to separate out particle size ranges, largely for the purpose of studying the in vivo and in vitro effects. Impactors also come in handy when size distribution is the objective because the initial sorting is done by the impactor sequence. In fact, simple impactors are even used to sort the incoming aerosol-laden air prior to capture in filters. When the tool is a portable personnel monitor, the target is almost always $PM_{2.5}$, and the impactor causes only that fraction to be collected.

Microscopy

This is the only size measurement technique that will be described. In many ways, it is the gold standard if the labor and time intensity is tolerable. The samples may be acquired from any sorting step, the most common being filters. The particles are then prepared differently for each technique. These range from simple low-resolution optimal microscopy, to higher-resolution optical means, to electron microscopy. Optical methods are limited to resolution of about 200 nm and magnification of about 2000 ×. Electrons have a wavelength smaller by a factor of 10^5 than visible light. Consequently, electron microscopes can achieve magnification of 10^7, yielding a resolution of 50 picometers (10^{-12} m). But specimen preparation can be cumbersome, and the equipment is expensive.

Electron microscopes are of two types. The original invention, the transmission electron microscope (TEM) is still in use, although the newer scanning electron microscope (SEM) is more common. Both use high-energy electron beams focused onto the specimen. In TEM, the beam transmits through the specimen (hence the name), and the image is produced on the other side. This limits the thickness of the sample to about 100 nm. For thicker samples, slices are taken. This relatively onerous sample preparation is one factor limiting TEM use.

SEMs are much more forgiving on sample size and preparation. The sample is scanned in a rectangular form, known as raster scanning. This allows examination of a swath along the sample. The incident radiation on the specimen loses energy, the loss being converted into forms of scattered low- and high-energy electrons and X-rays. The change in intensity in any of these forms of energy creates the image required. An important variant is the one using X-rays, known as energy-dispersive X-ray spectroscopy (EDX, pronounced "edex"). Equipment with this feature is known as SEM with EDX attachment. The number and energy of the X-rays produced are determined by an energy-dispersive spectrometer, part of the EDX attachment. A typical representation of the result is with energy on the horizontal axis and counts estimated by the heights of the peaks at each of the energy locations for the constituent elements.

The EDX attachment is particularly useful in PM evaluation because the different oxidative states of the organic layer in a secondary organic aerosol are best identified, at least preliminarily, by this method. This information will inform further evaluation using in vivo or in vitro testing of toxic effects. When unexpected reductions in solar panel efficiency were observed from PM deposits (Chapter 13), the composition of the deposit informed the likely mechanism of action. In general, microscopy is a valuable tool for mechanistic understanding, but the time and labor involved limits the scope of use. As with any investigative method, a combination of tools is most useful.

Satellite monitoring

Satellite-based estimation of PM overcomes current limitations of terrestrial monitoring stations in terms of cost-effective deployment, especially in sparsely populated areas.[10] The basis of the technique is that $PM_{2.5}$ (or PM_{10}) can be estimated from satellite-derived aerosol optical depth (AOD). It is the extinction of sunlight in specific wavelengths by aerosols in the entire column of the atmosphere.

$$\mathrm{AOD} = \mathrm{PM}_{2.5}\, H\, S, \tag{6.2}$$

where H is the well-mixed boundary layer of height H, with no overlying aerosols and having optical properties that are similar, and S is the specific extinction coefficient ($m^2\ g^{-1}$) of the aerosol at ambient relative humidity. The term S comprises a formula containing the ratio of ambient and dry extinction coefficients, aerosol mass density, aerosol effective radius, and the Mie extinction efficiency. The relationship is not reproduced here, but suffice to say each parameter has considerable uncertainty, as does adherence to the definition of H. As discussed later in this chapter, these uncertainties are in part resolved by machine-learning techniques and combining select terrestrial measurements. The real-time aspect of the satellite-based measurement is a key; the results are available within 3 h of the satellite overpass and can be used to provide forecasts and alerts.

The greatest promise lies in combining satellite imagery with ground-based monitoring. However, this applies to any significant degree only to the United States, Europe, and China. Even in these areas, existing monitoring stations are still few and tend to be concentrated in industrial regions or areas with heavy traffic. However, the increasing availability of low-cost sensors is likely to improve the situation. One would expect the satellite-based modeling to drive the needs for the most appropriate ground monitoring.

A recent publication by Donkelaar et al.[11] combines multiple satellite imagery products with ground-based monitoring, where available and overlays statistical methodology based on regression. Their results on $PM_{2.5}$ distribution compared very favorably ($R^2 = 0.81$) with out-of-sample cross-validated data from ground-based monitors. Their 2010 results show that worldwide population-weighted annual average $PM_{2.5}$ concentration is 32.6 $\mu g/m^3$, when compared with the World Health Organization's guidelines of 10 $\mu g/m^3$. This tripling over the guidelines is mostly due to contributions from Africa and Asia. Seven of the most polluted cities in the world are in India, with annual averages exceeding 100 $\mu g/m^3$. The principal takeaway from satellite imagery enhanced with ground monitoring and aided by data analytics is that large spatial conclusions can be drawn, which would be impractical with ground monitoring alone. Furthermore, transmigration across continents can be detected and quantitated to a degree. This helps underline the problem of particulate emissions as being global rather than local.

Gupta et al.[12] studied the impact of California wildfires on regional air quality. They point out that wildfire PM, particularly $PM_{2.5}$, can have a profound impact on regional health, despite being episodic (in contrast to automotive PM that is more nearly constant). For the purpose of this discussion, the primary contribution of the paper is the combining of satellite-derived observations with data from low cost air quality monitors (LCAQMs), which were discussed earlier in this chapter. They compared EPA data conforming to the FEM with the LCAQM data. The latter were then combined with satellite-based estimation of airborne aerosols during a wildfire event in October 2017. Interestingly, the ground-based information was captured by citizen science groups and by regulatory agencies. Low cost portable monitors have always held the promise of high granularity in measurement and the possibility of a volunteer citizen force to collect data. The issue has been the low quality and whether this aspect could be overcome by neural networks analyses and combination with other forms of data. The paper appears to conclude that, at least in this instance, good results can be obtained with the combination. AOD $PM_{2.5}$ correlations work best when the PM is concentrated at the bottom of the column and can approach $R^2 = 0.75$. However, in the case of wildfires, the assumption of this concentration does not hold, and additional ground-based data are needed. The study concludes that the combination of LCAQM data with satellite-based aerosol investigation is good enough to be actionable without the use of relatively expensive and sparse standard EPA monitoring stations.

Wang et al.[13] used the combination of satellite measurement with ground-based measurements to estimate PM_1, as opposed to the larger particles in the studies already described. They found high correlations ($R^2 = 0.8$) with conventional measurements. The model performed better in the afternoon than in the morning. It also was more effective at lower pollution levels. Song et al.[14] quantified the already known shortcoming of AOD in cloudy conditions and suggested remedies.

Satellite-based PM estimation is certainly important, in large measure because of the coverage area and timeliness for meteorological alerts. But the relatively new methods using ground-based monitoring in combination are important advances. They are made more feasible by the increased prevalence of low-cost monitoring devices. An interesting twist is the use of the public through citizen science approaches.[12]

Close-up: Living walls

Trees and shrubbery have long been used as barriers to the nuisance of road traffic. Fig. 6.4 shows a Chinese juniper with small branched needles as an example. Visual isolation and sound reduction are the objectives. With the increased recognition of health effects of traffic-originated PM, studies have commenced to quantify the PM-capturing capability of plant matter. To a degree, the foliage acts as a sensor of PM that deposits on leaves. These deposits are examined by the techniques described in this chapter. SEM with EDX identifies the species.[15] Ordinarily, high-intensity electron beams are not used on living matter, due to the propensity for damage. But here the researchers are examining the deposits and not the leaves. Damage to the leaves is irrelevant. SEM-EDX identified the elemental distribution of the compounds on the leaves as consistent with that of PM from vehicle exhaust. This is an example where the precise distribution of molecules is not needed to confirm provenance. Before the elimination of tetraethyl lead (an octane enhancer) from gasoline, its mere presence would have proven as a marker. But the EDX analysis has a level of sophistication that does not need the presence or absence of a species to establish provenance. In fact, Weerakkody et al. made a point of the absence of lead due

Fig. 6.4 *Juniperus chinensis* showing the branched needle character.

to the move away from lead in gasoline. Earlier studies had used lead as a marker of automotive sourcing, and these authors note why it is absent in their data.

Living walls are distinguished from façade architecture that usually uses climber species such as ivy. All foliage is not equally adept at capturing PM species. Living walls use tree-like shrubs, as opposed to creepers. These are planted in irrigated and fertilized containers, usually rectangular and placed parallel to the road. The Weerakkody study is among the most recent and one of the few that target roadside measurements. The planter is movable to different locations. Such a system has a narrow depth when compared with a bank of trees and is much more amenable to urban usage.

Evergreen conifers had been identified in previous studies as more efficient in capturing particulates. The Weerakkody study confirmed that noting that the greatest density of PM was found on *J. chinensis*. Every particle size distribution had this characteristic. $PM_{2.5}$ was most striking, with a density on *J. chinensis* more than double that on its nearest neighbor, *Veronica vernicosa*, which is also an evergreen but a shrub with small elliptical, glossy, leathery leaves. *V. vernicosa* was the runner-up in every size category. Several other evergreens with small leaves did not perform nearly as well. One reason for the better performance by small branched needles, when compared with flat leaves, could simply be the surface area-to-volume ratio. One would expect the small branched needle morphology to have more surface area than the flat leaves per the 100 cm^2 area in the study.

This book has emphasized numbers of particles as opposed to mass per unit volume of air. If we accept the premise that the smaller particles are the worse actors, then another observation in the Weerakkody et al. study is on point. They noted that, for all plants tested, the numbers of particles captured were greater in the smaller ranges. For example, the two leading species in terms of efficiency of capture, *J. chinensis* and *V. vernicosa*, had mean capture figures detailed in Table 6.2.

Preliminary evidence exists for the ability of rain to wash off PM from the leaves. If this is confirmed for the most promising species, living walls offer the promise of being regenerated after initial loading. Reliance on rain may not be needed if periodic sprinkling is built into the system.

Table 6.2 Particulate matter capture on a 100-cm^2 living wall, by species and particle size.

	Mean number of particles × 10^6		
Species	**PM_{10}**	**$PM_{2.5}$**	**PM_1**
Juniperus chinensis	200	490	4800
Veronica vernicosa	100	190	2400

Values approximated from Fig. 2 of Weerakkody et al.

The best species being evergreen has the obvious benefit of yearlong coverage. Also, a dense growth species such as juniper provides a visual barrier and to some extent a sound barrier. Communities would do well to invest in juniper hedges as roadside shrubbery. If Junipero Serra Boulevard near Stanford University and other locations in the South Bay were to be so festooned, the road name could take a different meaning. Saint Junípero Serra could well be smiling in approval from on high.

References

1. May KR. The Cascade impactor: an instrument for sampling coarse aerosols. *J Sci Instrum*. 1945;22:187–195.
2. Chartier R, Phillips M, Mosquin P, et al. A comparative study of human exposures to household air pollution from commonly used cookstoves in Sri Lanka. *Indoor Air*. 2017;27:147–159. https://doi.org/10.1111/ina.12281.
3. Bulot FMJ, Johnston SJ, Basford PJ, et al. Long-term field comparison of multiple low-cost particulate matter sensors in an outdoor urban environment. *Sci Rep*. 2019;9 (1):7497. https://doi.org/10.1038/s41598-019-43716-3.
4. Pillarisetti A, Carter E, Rajkumar S, et al. Measuring personal exposure to fine particulate matter ($PM_{2.5}$) among rural Honduran women: a field evaluation of the ultrasonic personal aerosol sampler (UPAS). *Environ Int*. 2019;123:50–53. https://doi.org/10.1016/j.envint.2018.11.014.
5. Qin X, Xian X, Deng Y, et al. Micro quartz tuning fork based $PM_{2.5}$ sensor for personal exposure monitoring. *IEEE Sensors J*. 2018. https://doi.org/10.1109/JSEN.2018.2886888.
6. Fisher JA, Friesen MC, Kim S, et al. Sources of variability in real-time monitoring data for fine particulate matter: comparability of three wearable monitors in an urban setting. *Environ Sci Technol Lett*. 2019;6(4):222–227. https://doi.org/10.1021/acs.estlett.9b00115.
7. Bennett JE, Tamura-Wicks H, Parks RM, et al. Particulate matter air pollution and national and county life expectancy loss in the USA: a spatiotemporal analysis. *PLoS Med*. 2019;16(7). https://doi.org/10.1371/journal.pmed.1002856.
8. Hansen ADA, Rosen H, Novakov T. Real-time measurement of the absorption coefficient of aerosol particles. *Appl Optics*. 1982;21(17):3060–3062. https://doi.org/10.1364/AO.21.003060.
9. Amaral, S. S., Carvalho, J. A. de, Costa, M. A. M., & Pinheiro, C. (2015). An Overview of Particulate Matter Measurement Instruments. doi:https://doi.org/10.3390/atmos6091327.
10. Mei L, Strandgren J, Rozanov V, Vountas M, Burrows JP, Wang Y. A study of the impact of spatial resolution on the estimation of particle matter concentration from the aerosol optical depth retrieved from satellite observations. *Int J Remote Sens*. 2019;40(18):7084–7112. https://doi.org/10.1080/01431161.2019.1601279.
11. van Donkelaar A, Martin RV, Brauer M, Hsu NC, Kahn RA, Levy RC. Global estimates of fine particulate matter using a combined geophysical-statistical method with information from satellites, models, and monitors. *Environ Sci Technol*. 2016;50 (7):3762–3772.
12. Gupta P, Doraiswamy P, Levy R, et al. Impact of California fires on local and regional air quality: the role of a low-cost sensor network and satellite observations. *GeoHealth*. 2018. https://doi.org/10.1029/2018GH000136.

13. Wang W, Mao F, Zou B, et al. Two-stage model for estimating the spatiotemporal distribution of hourly PM1.0 concentrations over central and East China. *Sci Total Environ*. 2019;675:658–666. https://doi.org/10.1016/j.scitotenv.2019.04.134.
14. Song Z, Fu D, Zhang X, et al. MODIS AOD sampling rate and its effect on $PM_{2.5}$ estimation in North China. Atmospheric Environment, United States Environmental Protection Agency. (2016, August 11). In: *Evaluation of Emerging Air Pollution Sensor Performance [Overviews and Factsheets]*; 2019. Retrieved June 26, 2019, from US EPA website: https://www.epa.gov/air-sensor-toolbox/evaluation-emerging-air-pollution sensor-performance.
15. Weerakkody U, Dover JW, Mitchell P, Reiling K. Quantification of the traffic-generated particulate matter capture by plant species in a living wall and evaluation of the important leaf characteristics. *Sci Total Environ*. 2018;635:1012–1024. https://doi.org/10.1016/j.scitotenv.2018.04.106.

CHAPTER SEVEN

Estimation of toxicity of airborne particulates

Epidemiological studies have shown that exposure to particulate matter (PM) is strongly correlated with hundreds of thousands of deaths a year, but many challenges remain in understanding which features of PM exposure are the drivers of mortality.[1,2] Different health outcomes were observed in different regions of the United States for the same PM mass increase, suggesting that mass may not be the prime driver of mortality.[3–5] Laboratory-based toxicological studies are unable to identify this driver and often show few health effects from the concentration levels found in ambient air.[6,7] Part of the challenge in unraveling this puzzle is that few toxicological studies are conducted to mimic realistic human inhalation or use mixtures representative of those found in ambient air. PM is coemitted with a variety of gases, and they are constantly interacting (Fig. 7.1). Additionally, PM continues to interact with gases present from other sources and can undergo compositional changes influenced by sunlight. Ambient PM and its dynamic complexity have proven to be a challenge to reproduce in toxicological studies.

In vitro and in vivo studies

Studies of the toxicological impact of PM on human organs fall broadly into two categories. In vivo methods address effects that directly target mammalian organs such as the lungs. In most instances, animal models are used first, with mice and rats being the most common. These studies are often followed by confirmation in human subjects. In vitro methods (literally meaning in glass) address the impact of agents on mammalian tissue. These bench laboratory methods are well suited for studying multiple variables and deciphering the underlying mechanisms.

Limited studies were conducted on the combined toxicological effects of PM and gases, their interaction with other pollutants, and the influence of atmospheric oxidation.[8] This is in part due to how in vitro exposures have been done to date. Most toxicological studies are limited to assessing

Particulates Matter
https://doi.org/10.1016/B978-0-12-816904-9.00004-0

Fig. 7.1 Aerosols come from different sources that result in exposures of vastly different compositions of chemicals to humans. Even the composition of this ink harvested from atmospheric aerosols could vary depending on where it was collected. *(Courtesy graviky labs, Inc.)*

gas-phase or PM exposure separately and, thus, miss any synergistic influences on toxicity.[9,10] Furthermore, these studies commonly use freshly generated particles from a single source (e.g., a diesel engine)[11] or concentrated ambient PM.[12] The oxidation products due to atmospheric chemistry are also not commonly included. Although these approaches have generated valuable data, they do not mimic the dynamic nature of exposure experienced in the ambient air. The complexity of gases and PM found in the atmosphere requires a new approach that includes not only the freshly emitted mixtures but also products of atmospheric oxidation. These experiments must be able to also isolate exposure to the individual constituents of the complex mixtures so that their relative toxicities can be assessed and compared with each other and to the whole mixture. Even with these improved capabilities in generating accurate repeatable in vitro exposures, challenges still exist in mimicking realistic human inhalation of PM.

New studies are needed to systematically disentangle the drivers of toxicity of the complex exposures of PM found in ambient air. In vivo techniques are most physiologically appropriate to quantify biological responses. They use an exposure chamber with mice or rats. Although the National Academy of Sciences states that animal models are valuable and the results can be extrapolated with care to human in vitro results,[13] there are significant barriers in their use in toxicology. They are expensive, sometimes prohibitively so for systematic evaluations such as disentangling the drivers of PM toxicity. Organizations such as the National Academy of Sciences have led movements to reduce animal use in toxicology.[14] Also, there are inherent limitations on extrapolating from animals to humans.[15] These barriers are significant and are the reason the scientific community has settled on in vitro inhalation assessment. These techniques use variants in the transport of pollutants and apply them to targeted biological models. The models can range from the use of cell-free biochemical assays[16] to cells that are part of a monoculture system.[17–19] In the recent years, biological models have come to include multicell cultures grown at the air-liquid interface (ALI) and cells from human donors.[20–22]

Despite the realism of the latest models using biomarker data to understand biological mechanisms,[23,24] comparing those data with results generated in vivo remains challenging. Several in vivo studies have found that biomarker results did not correlate with those found in vitro.[25] Recently, researchers showed that NF-kB activation in vitro predicted human neurotoxicity better than in vivo mouse methods.[26] There is also a strong evidence of in vitro and in vivo correlation. This includes studies assessing exposure to c-BC-Pb[27] and skin exposure to PM.[28] The uncertainty about comparability and utility of data lies in the accuracy of exposure and the method of transport of pollutants to the targeted biological model. In vitro techniques that are better indicators of in vivo animal and human toxicity are needed. The solution may lie in recent advancements in in-vitro technologies that use a more realistic inhalation model.

Filtrate resuspension exposure method

The value of new in vitro technologies is best appreciated by considering the details of the resuspension method, the conventional means for assessment of PM toxicity (Fig. 7.2, left and middle panels). The PM is collected over the course of several hours onto filters. The particles are extracted with a solvent into a resuspension liquid and applied to epithelial

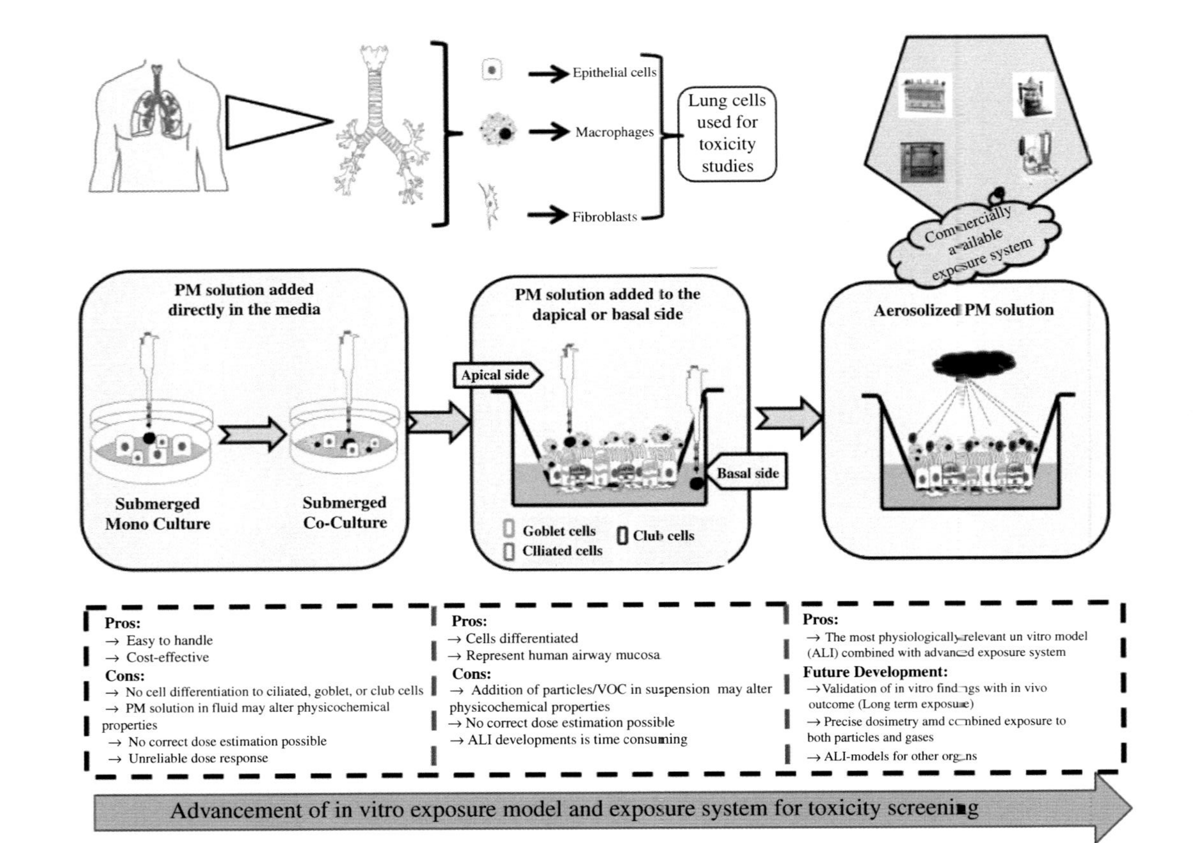

Fig. 7.2 Three types of in vitro exposure models for assessing the effects of PM on human lung cells. In the left and middle panels, the PM is resuspended in liquid and applied to the cells. In the right panel, the PM is aerosolized and deposited onto the ALI without using the resuspension method.[29]

lung cells by submerging them for a specified period, after which toxicity is assessed. The method has several advantages, including ease of use and relatively low cost. Assessment can be done on filters that have been archived or shipped from remote locations. Nevertheless, there are several shortcomings of the resuspension method that limit it for disentangling the drivers of PM toxicity.

The resuspension method is the most common assessment technique used in the toxicological field. Despite its widespread use, its limitations must be appreciated to determine the most suitable use. The two critical shortcomings are that it does not mimic real human inhalation and the collection method alters the size and composition of the PM from its original state in the atmosphere. The latter occurs from the moment the particles are collected on a filter, where they are continually being exposed to the sample flow and, thus, are no longer in equilibrium with surrounding ambient gases. Likewise, once eluted by the resuspension fluid, the PM is no longer with any gases, and the composition is changed by the solvent. Particles in the resuspended liquid agglomerate, especially the smallest ones, hampering analysis for size and the critical question of the importance of the size to toxicity. Because diffusion in liquids is slower than in air, the PM in liquid needs to be at a higher concentration to ensure interaction with the targeted biological model. This results in much larger exposure concentrations than are found in the ambient environment. This alteration prevents the evaluation of the components of PM and gases found in ambient air. These known shortcomings are being addressed with in vitro techniques that make use of the ALI.

Air-liquid interface exposure method

Exposure of cultured cells at the ALI, as shown in Fig. 7.2, is an air-to-cell condition that can be regarded as an effective surrogate for inhalation, more realistic than resuspension. Cells are cultured on a porous membrane that allows direct air-cell interaction on the apical surface, whereas the culture medium on the basolateral side provides nutrients for viability. ALI is now the gold standard for in vitro exposure.[30] ALI is more sensitive than resuspension and can show a significant inflammatory response to PM that resuspension cannot do. In the investigation by Lichtveld et al., diesel engine emissions were combined with an urban-like, complex volatile organic carbon compound mixture and injected into the rooftop chamber shown in Fig. 7.3.[31] The injected mixture was modified by exposure to natural light

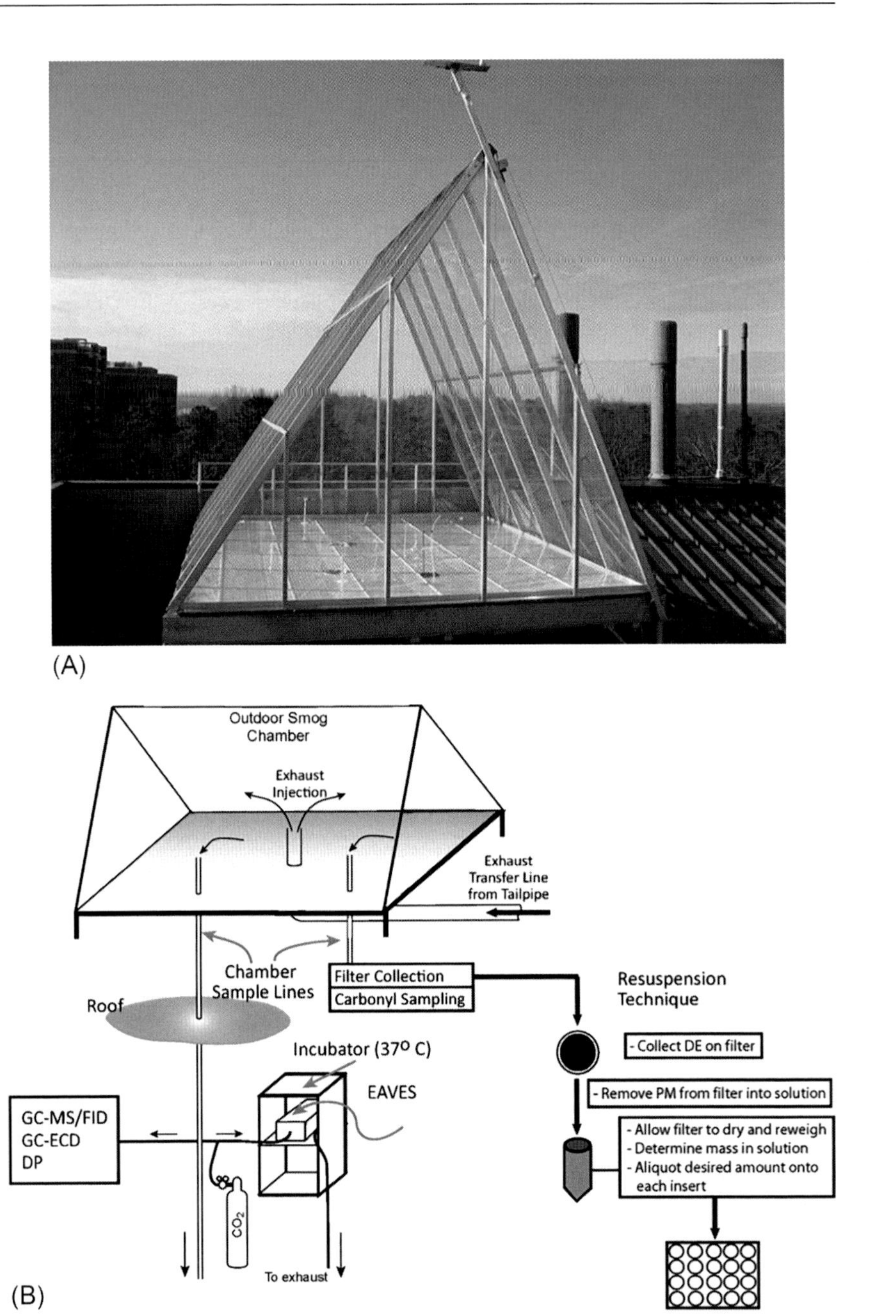

Fig. 7.3 The University of North Carolina rooftop chamber (A) and a schematic of the chamber, the laboratory sampling systems, and the electrostatic aerosol in vitro exposure system (B), an earlier prototype of the Gillings and CelTox instruments. PM was collected by a filter at the end of the duct through the floor of the chamber and resuspended for analysis.[31]

from sunrise to sunset. Filters were collected before sunrise and after sunset, and toxicity was assessed by resuspension. At the same time as the filter collection, direct deposition of PM was accomplished at the ALI using the electrostatic aerosol in vitro exposure system (EAVES), an earlier prototype of the Gillings and CelTox instruments. The setup provided the data needed to compare this novel technology with established resuspension methods.

This study was the first to determine whether filter-collected PM would affect the toxicity estimation of PM differently when compared with direct exposure. Chemical measurements during the photochemical aging process assured that primary hydrocarbons were consumed and that a large variety of gas-phase carbonyls and other oxidized products were generated, many of which were likely to be partitioned into the PM. The same analysis of resuspended PM, however, showed that no carbonyls or other oxidized organics were present, other than in the water and media blanks. This is chemical evidence that filter-collected and resuspended PM has lost potential key drivers of cellular toxicity. Direct exposure using the EAVES elicited a significant genomic response, as shown in the left panel of Fig. 7.4.[31] The right panel shows that the resuspension method failed to induce a positive response from cells at concentrations equal to those exposed by the EAVES. Only when the concentration of PM was increased 16-fold was there a significant response (not shown), and it was only approximately 50% of that for the tissue being exposed in the EAVES.

Despite the promise of a more realistic biological exposure through ALI, the best method to transport the pollutant to that interface is still uncertain, and comparative evaluations of performance of various systems designed to transport particles and gases at the ALI are lacking. A recent review of ALI approaches found that in over 81 publications, about half did not use resuspension methods for exposure. Of those, only six compared their results to resuspension.[32] Even among these six, there is a lack of consistency among exposure conditions, biomarkers used, and characterization of the gas and particle exposure. This makes comparative assessments difficult and prevents the development of best design practices in using these methods.

In an attempt to fill the void, the US Environmental Protection Agency (EPA) assessed five ALI methods for effectiveness in delivering particles and gases onto cell culture inserts. Table 7.1 lists them: two developed by the University of North Carolina at Chapel Hill (UNC-CH), one by the EPA, and two commercially available instruments produced by Vitrocell.[34] The gas in vitro exposure system relies on diffusion of gases for ALI exposure and was successfully deployed in the field.[35] UNC-CH also developed the

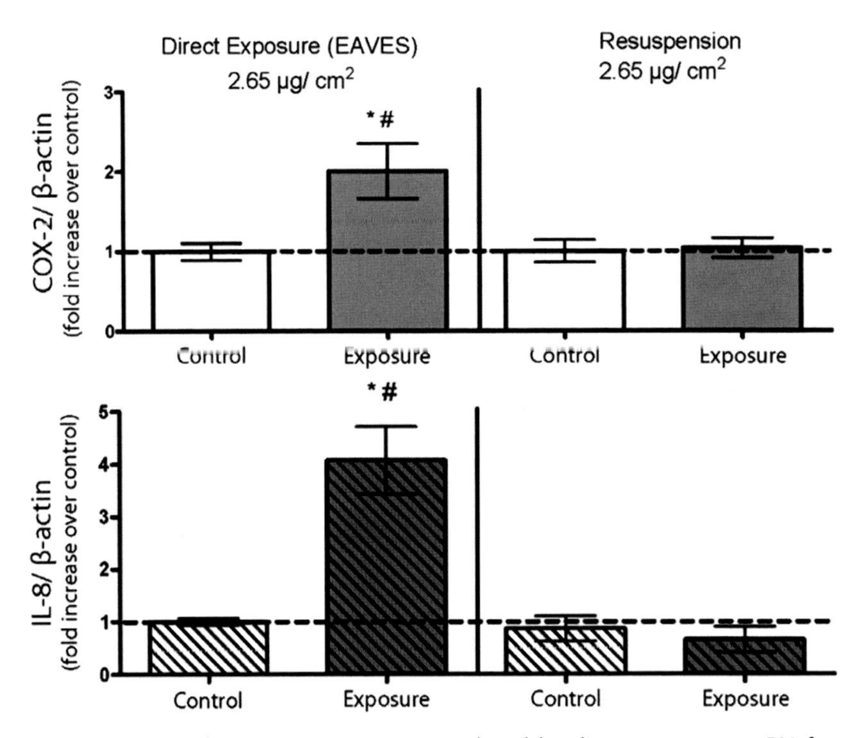

Fig. 7.4 COX-2 and IL-8 m-RNA expression induced by direct exposure to PM from photochemically aged diesel exhaust and urban gases transported by EAVES (left) and resuspension (right). With 2.65 $\mu g/m^2$ of exposure only, EAVES detected a significant increase. *Indicates a difference from nonexposed incubator control, # indicates a difference from resuspension, $P < 0.05$. Error bars represent the mean $\pm$ standard error of the mean.

Table 7.1 In vitro systems in the US EPA comparative analysis.[33]

Manufacturer	System's name	Principle of operation	
		Gases	**Particles**
UNC-Chapel Hill	Gillings	Diffusion	Electrostatic precipitation
UNC-Chapel Hill	GIVES	Diffusion	Diffusion or sedimentation
US EPA	CCES	Diffusion	Thermophoresis, diffusion or sedimentation
Vitrocell	6/3 CF	Diffusion	Electrostatic precipitation
Vitrocell	6 CF	Diffusion	Diffusion or sedimentation

CCES, cell culture exposure system; *GIVES*, gas in vitro exposure system; *USEPA*, United States Environmental Protection Agency.

Gillings' Sampler that contains all necessary components to ensure cell viability and uses a corona wire to electrostatically charge the particles. An electric field is applied above the cells to drive the PM directly onto the cells that are cultured at the ALI. The corona wire allows unipolar charging of particles to enhance deposition, and the particle loading is consistent regardless of size.[31] There are two models of the cell culture exposure system by the EPA. One accommodates a 6-well format and the other a 24-well format. The 6-well format has a thermophoretic capability that enhances deposition.[33] The Vitrocell 6 CF relies solely on diffusion or sedimentation, whereas the 6/3 CF uses the particle's natural electrical charge for electrostatic precipitation (ESP).

The instruments were subjected to identical testing protocols. They were exposed to 125 ppb of O_3 gas for 1 h. Instead of biological cells, an indigo dye-impregnated filter and a fluorescence-based analytical method were used to quantitate exposure. The effectiveness of particle deposition was determined using fluorescent polystyrene latex spheres with diameters of 50 nm and 1 μm as a surrogate for PM. They were nebulized and delivered onto all instruments containing filters inside each cell culture insert as a collection substrate. Deposition was quantified by dissolving the particles in ethyl acetate and using a spectrofluorometer. The Gillings' Sampler outperformed all the others in the delivery of gas-phase chemicals and particles of both diameters. Details of the study are being prepared for publication.

The study demonstrated that particles less than 1 μm in diameter pose a difficult challenge for these new systems if they rely solely on diffusion and sedimentation forces for deposition onto cell culture inserts. To enhance deposition for submicron particles, an external force such as thermophoresis or electrostatics is required. A new ESP-ALI has been developed to improve nano- and submicron-sized particle collection efficiency.[36] Initial testing shows that it can collect particles down to 300 nm with uniformity in deposition. To mitigate ozone generation, the researchers used a novel electrospray charging technique instead of a corona-discharging process. Wide availability of such instruments could enhance assessment of the toxicity of submicron PM.

A new commercial ALI device

Several new technologies have emerged that provide direct exposure of cultured airway cells at the ALI, allowing complex air pollution mixtures to be tested in their ambient state. Although most of these devices were

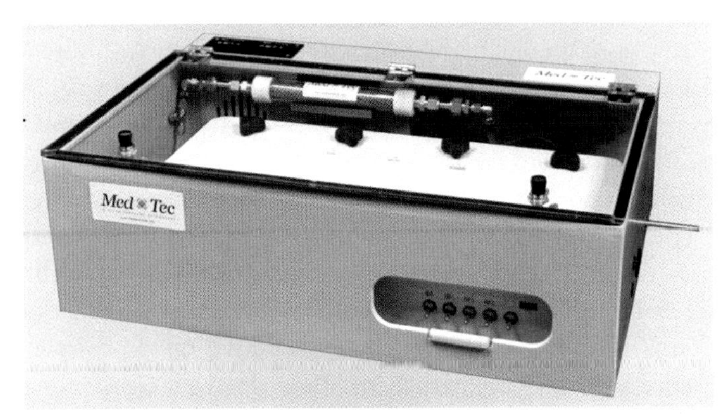

Fig. 7.5 MedTec CelTox Sampler for exposing cultured cells to gases and aerosols at the ALI. (http://www.medtecbiolab.com).

developed in research laboratories, a few are commercially available. One, based on the Gillings' Sampler, is the CelTox Sampler from MedTec (Fig. 7.5). It tests gases and aerosols. It has a custom-designed, reusable multiwell plate that holds up to six 30-mm diameter Millicell-CM cell culture inserts or 12 12-mm Snapwell inserts. Other formats using standard disposable multiwell plates can also be accommodated. The system can also use disposable multiwell plates for the miniaturized Ames assay in 6, 12, and 24 well formats. It is temperature and humidity regulated, and, to maintain cell viability, the sampled air is conditioned (warmed and moistened) prior to reaching the cell cultures. This is critical for robust biomarker data. The humidification system can add moisture to the sampled atmosphere without diluting the test atmosphere. The instrument contains a unipolar particle-charging source that can be turned off to rely on the particles' natural charge or turned on to charge the particles for deposition enhancement. The instrument operates under a vacuum that protects the user from unintentional exposure to the test atmosphere.

The CelTox Sampler has a small footprint, allowing it to sit on a cart or laboratory bench or inside a chemical hood. The superior performance of the Gillings' Sampler (Fig. 7.4) combined with ease of use provides a measurement platform with the potential to change how the research community quantifies in vitro aerosol toxicity. The sensitive and realistic ALI method enables the study of true biological and chemical impacts. The small footprint could make this instrument the standard for field work, although the deployment of live tissue presents a logistical challenge. Ultimately, an improved resuspension method and an ALI method will be effectively combined for field measurements, leaving the ALI as the preferred laboratory method.

Toxicity of biomass-sourced particulates: A case study

Airborne PM is responsible for 3.7–4.8 million premature deaths annually worldwide,[37] 92% of which occur in low- and middle-income countries.[38–40] Exposure to wood-burning cookstove emissions causes 2.8 million of these deaths.[41] About 66% of the population in India relies on solid biomass for cooking, whereas it is about 80% in subSaharan Africa.[42] Emissions from cookstoves consist of PM, various volatile organic compounds, and other gases that are harmful to health. These emissions, when released into the ambient environment, can undergo photochemical transformations. Understanding the roles that photochemistry and PM-to-gas composition have on toxicity requires the ability to expose human airway cells to photochemically aged and fresh pollutant mixtures and quantify the effects. A recent study used the CelTox to quantify that toxicity.

In practice, the imprecise nature of fuel type and condition and of air flow control in cookstoves determine burn conditions (e.g., smoldering versus flaming) that greatly affect emission characteristics. Combustion efficiency depends on stove design, fuel type, and fuel moisture content (fresh or dry), among other factors and has been found to vary between 80% and 96%.[43,44] The adverse health effects of cookstove emissions are recognized, and studies have linked exposure to PM with premature mortalities. The importance of the chemical fingerprint, photochemical aging, and physiochemical properties of PM in biological responses is still not well understood. A systematic investigation of the interaction between well-characterized emissions and biological systems is needed to generate comprehensive exposure-response profiles and provide guidelines for mitigating health risks.

A recent experiment has provided new insight into the most important toxicological drivers in woody mass combustion. It used a chamber specifically designed for biomass burning at North Carolina A&T State University.

The chamber consists of an approximately 9-m^3 reactor with a tube furnace in a temperature-controlled room. Optical properties were measured in conjunction with other analytical instruments, including CO and CO_2 monitors. Photochemical aging was achieved by ultraviolet lamps placed on two opposite sides of the chamber. The outputs, including the water-soluble

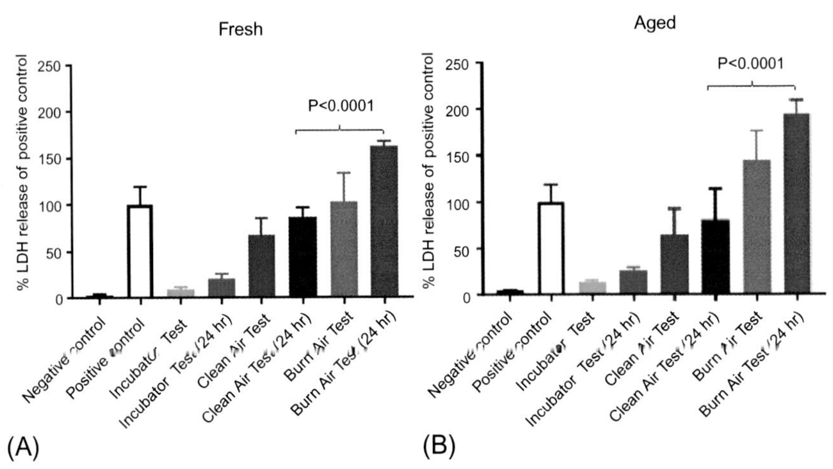

Fig. 7.6 Preliminary cytotoxicity results in epithelial lung cells from exposure to fresh (A) and aged (B) emissions of biomass burning. Data are normalized to the positive control. LDH is lactate dehydrogenase.

organic gases, water-soluble organic particles, and organic and elemental carbon, were fully characterized by a range of standard techniques, including ion chromatography and mass spectrometry. As a result, the researchers had precise knowledge of the size of the particles and the chemical differences arising from photochemical aging. The CelTox Sampler was used for ALI exposure of the BEAS-2B human lung epithelial cell line (ATCC CRL-9609).[45] Exposure occurred in real time for both fresh and aged biomass combustion emissions. The biological change was evaluated using two commercial assay toxicity kits for lactate dehydrogenase release (LDH assay, Sigma TOX-7) and gene expression analysis, building on the previous work published by UNC-CH.[35]

Fig. 7.6 shows preliminary results for cytotoxicity from these fresh and aged exposures normalized to the positive control. The positive control was epithelial lung cells exposed to 1% Triton X-100 for 2 h. There was a significant difference between the clean air exposure and those lung cells that were exposed to either the fresh or aged exposure. More importantly, there was also a significant difference ($P < 0.0001$) from the fresh exposures to the aged exposures. This was confirmed by earlier results that demonstrated that photochemical changes result in differential in vitro genomic responses, suggesting a change in the overall toxicity.[31,46] For example, a study at UNC-CH showed that when diesel emissions were photochemically aged in an urban multipollutant mixture, they had a 5- to10-fold increase in cytotoxicity over freshly emitted pollutants.[31] These results are encouraging and should spur further genomic analysis to be correlated with chemical characterization. Using these proven chemical and biological characterization methods, we can begin to identify the specific constituents of PM most relevant to overall toxicity. Further studies such as this will provide the data needed to better understand the scientific drivers of health impacts from biomass combustion emissions, thereby laying the foundation to inform field studies that will drive intervention and policy changes.

Follow-on research can provide gene expression analysis from these exposures. The combination of the well characterized transformation of emissions with robust genomic expression generated by the novel biological exposure systems will permit new correlations to determine the most relevant environmental drivers. They could be specific oxidized organic components of PM or simply the nanosized portion of aerosols. Once these drivers are identified, they can then be sought out in other well characterized biomass combustion emission sources such as wildfires. If, for example, certain oxidized organics are found to drive toxicity, then new epidemiological studies and in vivo experiments could be done using those critical environmental exposure conditions. Furthermore, the analytical and modeling tools may need to be updated to measure and predict these species. Ultimately, such tools and insights will inform policy decisions to produce the most effective controls for protecting public health.

References

1. Pope CAR, Dockery DW. Health effects of fine particulate air pollution: lines that connect. *J Air Waste Manag Assoc.* 2006;56(6):709–742.
2. Valberg PA. Is PM more toxic than the sum of its parts? Risk-assessment toxicity factors vs. PM-mortality "effect functions". *Inhal Toxicol.* 2004;16(Suppl 1):19–29.
3. Baxter LK, Duvall RM, Sacks J. Examining the effects of air pollution composition on within region differences in PM 2.5 mortality risk estimates. *J Expo Sci Environ Epidemiol.* 2013;23(5):457–465.
4. Dominici F, Peng RD, Bell ML, et al. Fine particulate air pollution and hospital admission for cardiovascular and respiratory diseases. *JAMA.* 2006;295(10):1127–1134.
5. Pinault L, Tjepkema M, Crouse DL, et al. Risk estimates of mortality attributed to low concentrations of ambient fine particulate matter in the Canadian community health survey cohort. *Environ Health.* 2016;15(1):18.
6. Laks D, de Oliveira RC, de Andre PA, et al. Composition of diesel particles influences acute pulmonary toxicity: an experimental study in MICE. *Inhal Toxicol.* 2008;20(11):1037–1042.
7. Sunil VR, Patel KJ, Mainelis G, et al. Pulmonary effects of inhaled diesel exhaust in aged mice. *Toxicol Appl Pharmacol.* 2009;241(3):283–293.
8. Fuzzi S, Baltensperger U, Carslaw K, et al. Particulate matter, air quality and climate: lessons learned and future needs. *Atmosph Chem Phys.* 2015;15(14):8217–8299.

9. Farina F, Sancini G, Mantecca P, Gallinotti D, Camatini M, Palestini P. The acute toxic effects of particulate matter in mouse lung are related to size and season of collection. *Toxicol Lett.* 2011;202(3):209–217.
10. Steenhof M, Gosens I, Strak M, et al. In vitro toxicity of particulate matter (PM) collected at different sites in the Netherlands is associated with PM composition, size fraction and oxidative potential—the RAPTES project. *Part Fibre Toxicol.* 2011;8:26.
11. Schlesinger RB, Kunzli N, Hidy GM, Gotschi T, Jerrett M. The health relevance of ambient particulate matter characteristics: coherence of toxicological and epidemiological inferences. *Inhal Toxicol.* 2006;18(2):95–125.
12. Steerenberg PA, van Amelsvoort L, Lovik M, et al. Relation between sources of particulate air pollution and biological effect parameters in samples from four European cities: an exploratory study. *Inhal Toxicol.* 2006;18(5):333–346.
13. Zavala J, Freedman AN, Szilagyi JT, et al. New approach methods to evaluate health risks of air pollutants: critical design considerations for in vitro exposure testing. *Int J Environ Res Public Health.* 2020;17(6):2124.
14. Stephens ML, Goldberg AM, Rowan AN. The first forty years of the alternatives approach: refining, reducing, and replacing the use of laboratory animals. In: Salem DJ, Rowan AN, eds. The State of the Animals: 2001. Washington, DC: Humane Society; 2001:122–135.
15. Bracken MB. Why animal studies are often poor predictors of human reactions to exposure. *J R Soc Med.* 2009;102(3):120–122.
16. Mudway IS, Stenfors N, Duggan ST, et al. An in vitro and in vivo investigation of the effects of diesel exhaust on human airway lining fluid antioxidants. *Arch Biochem Biophys.* 2004;423(1):200–212.
17. Bhattacharya K, Andón FT, El-Sayed R, Fadeel B. Mechanisms of carbon nanotube-induced toxicity: focus on pulmonary inflammation. *Adv Drug Deliv Rev.* 2013;65 (15):2087–2097.
18. Gliga AR, Skoglund S, Wallinder IO, Fadeel B, Karlsson HL. Size-dependent cytotoxicity of silver nanoparticles in human lung cells: the role of cellular uptake, agglomeration and Ag release. *Part Fibre Toxicol.* 2014;11(1):11.
19. Nymark P, Catalán J, Suhonen S, et al. Genotoxicity of polyvinylpyrrolidone-coated silver nanoparticles in BEAS 2B cells. *Toxicology.* 2013;313(1):38–48.
20. Hiemstra PS, Grootaers G, van der Does AM, Krul CA, Kooter IM. Human lung epithelial cell cultures for analysis of inhaled toxicants: lessons learned and future directions. *Toxicol In Vitro.* 2018;47:137–146.
21. Lenz A-G, Karg E, Brendel E, et al. Inflammatory and oxidative stress responses of an alveolar epithelial cell line to airborne zinc oxide nanoparticles at the air-liquid interface: a comparison with conventional, submerged cell-culture conditions. *Biomed Res Int.* 2013.
22. Müller L, Brighton LE, Carson JL, Fischer II WA, Jaspers I. Culturing of human nasal epithelial cells at the air liquid interface. *JoVE (J Visual Exp).* 2013;80, e50646.
23. Clippinger AJ, Ahluwalia A, Allen D, et al. *Expert Consensus on an In Vitro Approach to Assess Pulmonary Fibrogenic Potential of Aerosolized Nanomaterials.* Springer; 2016.
24. Fröhlich E, Salar-Behzadi S. Toxicological assessment of inhaled nanoparticles: role of in vivo, ex vivo, in vitro, and in silico studies. *Int J Mol Sci.* 2014;15(3):4795–4822.
25. Breznan D, Karthikeyan S, Phaneuf M, et al. Development of an integrated approach for comparison of in vitro and in vivo responses to particulate matter. *Part Fibre Toxicol.* 2015;13(1):41.
26. Zhang H, Haghani A, Mousavi AH, et al. Cell-based assays that predict in vivo neurotoxicity of urban ambient nano-sized particulate matter. *Free Radic Biol Med.* 2019;145:33–41.
27. Jiang S, Shang M, Mu K, et al. In vitro and in vivo toxic effects and inflammatory responses induced by carboxylated black carbon-lead complex exposure. *Ecotoxicol Environ Saf.* 2018;165:484–494.

28. Piao MJ, Ahn MJ, Kang KA, et al. Particulate matter 2.5 damages skin cells by inducing oxidative stress, subcellular organelle dysfunction, and apoptosis. *Arch Toxicol.* 2018;92(6):2077–2091.
29. Upadhyay S, Palmberg L. Air-liquid interface: relevant in vitro models for investigating air pollutant-induced pulmonary toxicity. *Toxicol Sci.* 2018;164(1):21–30.
30. Zavala J, Greenan R, Krantz QT, et al. Regulating temperature and relative humidity in air–liquid interface in vitro systems eliminates cytotoxicity resulting from control air exposures. *Toxicol Res.* 2017;6(4):448–459.
31. Lichtveld K, Ebersviller S, Sexton K, Vizuete W, Jaspers I, Jeffries HE. In vitro exposures in diesel exhaust atmospheres: Resuspension of PM from filters verses direct deposition of PM from air. *Environ Sci Technol.* 2012;46(16):9062–9070.
32. Schmidt S, Altenburger R, Kühnel D. From the air to the water phase: implication for toxicity testing of combustion-derived particles. *Biomass Convers Biorefin.* 2019;9(1):213–225.
33. Zavala J, Higuchi MA. *Understanding Air-Liquid Interface Cell Exposure Systems: A Comprehensive Assessment of Various Systems under Identical Conditions.* San Antonio: Society of Toxicology Annual Meeting; 2018.
34. Vitrocell, 2002. Vitrocell (CULTEX systems). Retrieved Mar 19, 2002, from www.vitrocell.com.
35. Vizuete W, Sexton KG, Nguyen H, et al. From the field to the laboratory: air pollutant-induced genomic effects in lung cells. *Environ Health Insights.* 2015;9(Suppl. 4):15–23.
36. Hsiao T-C, Chuang H-C, Chen C-W, Cheng T-J, Chien Y-CC. Development and collection efficiency of an electrostatic precipitator for in-vitro toxicity studies of nano-and submicron-sized aerosols. *J Taiwan Inst Chem Eng.* 2017;72:1–9.
37. Cohen AJ, Brauer M, Burnett R, et al. Estimates and 25-year trends of the global burden of disease attributable to ambient air pollution: an analysis of data from the global burden of diseases study 2015. *The Lancet.* 2017;389(10082):1907–1918.
38. Heft-Neal S, Burney J, Bendavid E, Burke M. Robust relationship between air quality and infant mortality in Africa. *Nature.* 2018;559:254–258. https://doi.org/10.1038/s41586-018-0263-3.
39. Landrigan PJ, Fuller R, Acosta NJR, et al. The lancet commission on pollution and health. *The Lancet.* 2018;391(10119):462–512.
40. Lelieveld J, Haines A, Pozzer A. Age-dependent health risk from ambient air pollution: a modelling and data analysis of childhood mortality in middle-income and low-income countries. *Lancet Planet Health.* 2018;2(7):e292–e300.
41. Gordon SB, Bruce NG, Grigg J, et al. Respiratory risks from household air pollution in low and middle income countries. *Lancet Respir Med.* 2014;2(10):823–860.
42. IEA. *Energy and Air Pollution 2016—World Energy Outlook Special Report International Energy Agency.* Paris: France; 2016.
43. Coffey ER, Muvandimwe D, Hagar Y, et al. New emission factors and efficiencies from in-field measurements of traditional and improved Cookstoves and their potential implications. *Environ Sci Technol.* 2017;51(21):12508–12517.
44. Fleming LT, Weltman R, Yadav A, et al. Emissions from village cookstoves in Haryana, India and their potential impacts on air quality. *Atmos Chem Phys Discuss.* 2018;2018:1–21.
45. Stewart CE, Torr EE, Mohd Jamili NH, Bosquillon C, Sayers I. Evaluation of differentiated human bronchial epithelial cell culture Systems for Asthma Research. *J Allergy.* 2012;2012:11.
46. De Bruijne K, Ebersviller S, Sexton KG, et al. Design and testing of electrostatic aerosol in vitro exposure system (EAVES): an alternative exposure system for particles. *Inhal Toxicol.* 2009;21(2):91–101.

PART FOUR

Engineered interventions

The previous sections described how PM forms, the means for detection, health impacts, and techniques for assessing mechanisms of toxicity. This section gets into what one can do about it: engineered interventions. Fittingly, this is the longest section in the book. The interventions range from mere amelioration, such as by filtering the exhaust from combustion processes, to eliminating the source. An example of the last is displacing the conventional fuel being combusted with one more benign in the propensity to produce PM.

An entire chapter is devoted to electric vehicles, a transformational solution for tailpipe emissions, although attention must be paid to the source of the electricity. Wood-burning cookstoves are the single biggest global public health issue. Interventions have been largely ineffective, but certain fuel substitutions can be expected to succeed. Wildfires are statistically not a large contributor to mortality. But they get a full chapter in part because climate change is likely to increase their frequency and severity. On balance, this section is solutions oriented, with underlying optimism for better outcomes.

CHAPTER EIGHT

Wood fires: Domesticated

The combustion of wood for cooking and heating is the single largest contributor to airborne particulates. Some of the communities compelled to use wood for these purposes also indulge in burning household trash street side. The combustion conditions are quite different in these two modes, in part because the raw material is different. Then, there are the fires that are unintended: wildfires. As a component of a worldwide inventory, this last category is statistically small, the percentage of the total being in the low single digits. But the category merits discussion because the fires are devastating where they occur, because the contributory causes could well be anthropogenic and because the frequency is expected to increase substantially over the next decades. Wildfires will be covered in the next chapter; this chapter will deal simply with the intentional ones—domesticated, so to speak, as opposed to wild!

Wood combustion basics

Most wood burned voluntarily or involuntarily originates in forests. All tree trunks and limbs have essentially the same structure. The constituents are cellulose, hemicellulose, and lignin. Cellulose is the most prevalent, generally 40%–45% of the total mass.[1] It is a long-chain polymer of glucose, with alternate glucose units rotated by 180 degrees. This imparts a measure of rigidity. A high fraction of the cellulose is crystalline and will appear so in an X-ray diffraction image. Hemicellulose is a polymer of sugars other than glucose, such as mannose, galactose, and rhamnose. This polysaccharide is mixed with the cellulose, thus imparting more rigidity. But the greatest structural integrity comes from the third principal component, lignin. This is a cross-linked polymer with a high fraction of aromatic molecules. It fills the intercellular spaces between cellulose and hemicellulose. Importantly, it is covalently linked to the hemicellulose and consequently cross-links the polysaccharides. This imparts strength to the limb or trunk. The aromatics in lignin also give the smoke a distinctive smell. Smoked meats take advantage of this property.

Particulates Matter
https://doi.org/10.1016/B978-0-12-816904-9.00015-5

Whereas all woods have the character described above, species have the components in a different proportion. Softwoods have less hemicellulose and more lignin than hardwoods, the cellulose content being the same. This results in softwoods having higher carbon and lower oxygen in general. Consequently, hardwood combustion produces on an average 8600 BTU/lb., whereas softwood combustion produces on an average 9000 BTU/lb.[2] The soft and hard nomenclature is deceiving. The low specific gravity balsawood is softer than most woods but is classified as a hardwood. For combustion in stoves, hardwoods are generally preferred because they burn longer under the same conditions of air availability. The intrinsic oxygen content of wood is about 44%.[2] As discussed in Chapter 11, this is detrimental for pyrolytic oil production but is not a drawback for combustion as a fuel.

When burned, if complete combustion is achieved, the products are just CO_2 and water. Oxygen-deprived combustion will yield less energy and produce CO and volatile organics. Trash burning will often produce such conditions, sometimes referred to as smoldering. As discussed in earlier chapters, the combustion conditions determine the type and quantity of particulate matter (PM) produced. Oxygen-starved combustion is the means to charcoal, an important intermediate that is used for subsequent combustion.

Cookstove emissions

Except for some allusions almost in passing, most of our examination is of stoves used in low- and middle-income countries (LMICs). Wood burning in the higher income countries is prevalent, even as a primary means for cooking and heating, but the incidences are minor when compared with LMIC usage. More than 3 billion people use wood as the primary means for cooking.[3] The PM from this source alone is responsible for high morbidity that is difficult to estimate because of lack of reporting. The mortality figures range from 3.8 million to 4.3 million premature deaths attributed to PM, making this source the single biggest public health issue.[3]

LMIC cookstove usage is a way of life, and introduction of improvements encounters societal resistance. Aggressive fuel substitution of the type discussed later in this chapter is hampered by economics. Accordingly, the improvements to address reductions in PM are modest and in most cases are limited to indigenous changes to existing cookstoves. Most of the improvements are in two areas: the addition of an efficient chimney and better control of combustion parameters. The latter largely results in improved efficiency of the process. As will be reported below, testing to date shows modest improvement to the total mass of PM being produced per kilogram of fuel. This efficiency improvement does reduce the amount of fuel used for any particular purpose, causing the total PM from the purpose to be lower than when using original equipment. Improved chimneys can be expected to positively affect the health of the cook and in many instances the child in tow. But they will have negligible effects on the climate change mechanisms.

The commonest stoves are of stone or masonry with large mouths for feeding fuel and a top with an opening for a cooking utensil. The fuel is one of convenience, most commonly wood collected from forests in proximity. Cow dung is also used after drying in patties and comprises 10% of the fuel used in India.[4] The other two fuels are charcoal and coal.

For decades, investigations of the emissions from cookstoves have been conducted, mostly with a view to inform interventions. Design changes to improve efficiency have definite relevance. The divide appears when attempts are made to estimate the particulates produced in real-life situations. Laboratory studies in this space were shown to understate the expected field result. One study estimates differences at a factor of two to three.[5] This is not surprising because the same stove used in different countries and different parts of the same country for that matter can be expected to have different operating modes that may even change from meal to meal. An example would be the conditions needed for a high-heat fry versus a low-heat simmering stew. The fire conditions are different. The high-heat fry would be expected to have lower organic carbon (OC) than a smolder associated with a simmer.

One recent study by Rose Eilenberg and colleagues[6] will be discussed in some detail in part because it spans a variety of countries and largely because it is the first cookstove emissions study we encountered that measures and discusses particle count in addition to the usual mass-based observations. As noted throughout this book, particle count may prove to be a more relevant metric than mass per unit volume, especially when ultrafine particles are the principal species, which is the case with cookstove emissions.

The study was done in China, Honduras, Uganda, and India. The conditions were uncontrolled, in that no prescriptive equipment or methods were imposed. The measurements were made just above the stove or in the chimney if one was present. Particles were collected on filters and analyzed. The measured quantities included CO and CO_2, particle size distribution, elemental carbon (EC), and OC. The results were reported as emission factors (EF) that they define as

$$EF = C_p/(C_{CO} + C_{CO_2}) \times \text{fuel carbon content}, \quad (8.1)$$

where C_p is the background-corrected concentration of the pollutant, C_{CO} and C_{CO_2}, respectively, are the background-corrected concentrations of the produced CO and CO_2, and the fuel carbon content is the field-measured carbon content in the fuel, reported as kg C per kilogram of fuel.

The results are presented as $PM_{2.5}$ (particles with diameter less than 2.5 μm) EFs but with the finer granularity of EC and primary organic aerosols (POAs; Fig. 8.1). As described in Chapter 4, in this case the nuclei are EC, and the POA is the proportion of carbon nuclei that acquires an organic coating from the cogenerated volatiles from the organic matter present in the fuel. These species have very different radiative forcing. EC, also known as black carbon, is strongly absorbing in the infrared wavelengths and causes warming or positive radiative forcing (see Chapter 4 for definitions). Brown carbon, which is another term for POAs and secondary organic aerosols combined, is odd. It has positive radiative forcing in the upper atmosphere and negative closer to the surface of the earth, where the brown layer scatters

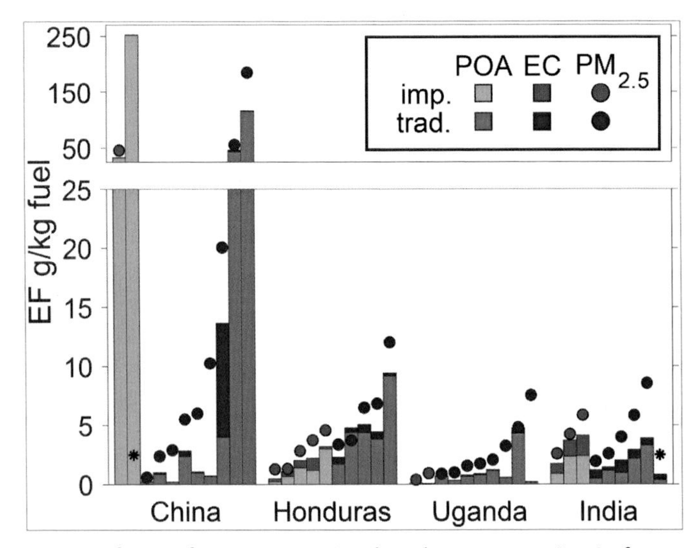

Fig. 8.1 Emission factors from unconstrained cookstove operation in four countries.[6] The green points and bars represent data from improved (imp.) cookstoves and the blue from traditional (trad.). The filled circles are for $PM_{2.5}$, and the bars are for particle detail. *(Courtesy copyright holder.)*

rather than absorbing the ultraviolet wavelengths. Their observations of the EC:POA ratio (they use EC:OC that in this case would be POA) are instructive. Although Eilenberg et al. do not explicitly discuss this, areas with high EC content could be expected to have a definite positive impact on warming. However, given the extraordinary impact of cookstove emissions on health, the effect on global warming is almost an afterthought. The total contribution to the atmosphere of both types of carbon from this source is believed to be 20%.[7] Removing all cookstove-based emissions is computed to generate from 0.16 K warming to 0.28 K cooling,[8] underlining again the uncertainty based on a host of burn parameters. Accordingly, for the rest of this chapter, we will focus on aspects that affect health, not climate change.

The China figures are largely from the combustion of coal and biomass pellets. Some of the highest emissions are from unprocessed coal (the most common processing is washing). Note the broken scale on the vertical axis to accommodate the high China values. The majority of the $PM_{2.5}$ is the combination of EC and POA. The lighter shades represent OC, or POA, and the darker shades are for EC. The lowest emissions are from Uganda, and these were almost all using charcoal as the feed. The authors caution that the emissions in the production of the charcoal were not accounted for.

The Honduras and India experiences were the most comparable because the sole fuel was wood, with similar properties. The differences underline the dangers in generalizing statistics and using them for broad conclusions. Just a visual interpretation points to higher EC proportions in India (compare the lighter and darker shades). Hotter fires with more complete combustion yield higher proportions of EC. Honduran cooking appears to have used smoldering fires. Another caution is that, in either country, regional variations in cuisine could cause different fire parameters.

The Eilenberg study also examines particle size distribution. Given the substantial differences in fuel types, the traces are not readily comparable. However, the Honduras and India experiences ought to be similar or understandably different. Fig. 8.2 shows distributions and particle counts. This last detail is important because, as discussed throughout the book and especially in Chapters 4 and 11, ultrafine particles may not have much mass but the particle count may be high. The Uganda experience would appear to exemplify this point. Note from Fig. 8.1 that the Uganda total $PM_{2.5}$ runs much lower than that in India, and yet the particle counts are over an order of magnitude higher than those in India, as seen in Fig. 8.2E. Note also that the Uganda particle distribution is skewed to smaller diameters than that of India, with the peaks about 60 nm apart. One would expect the Uganda

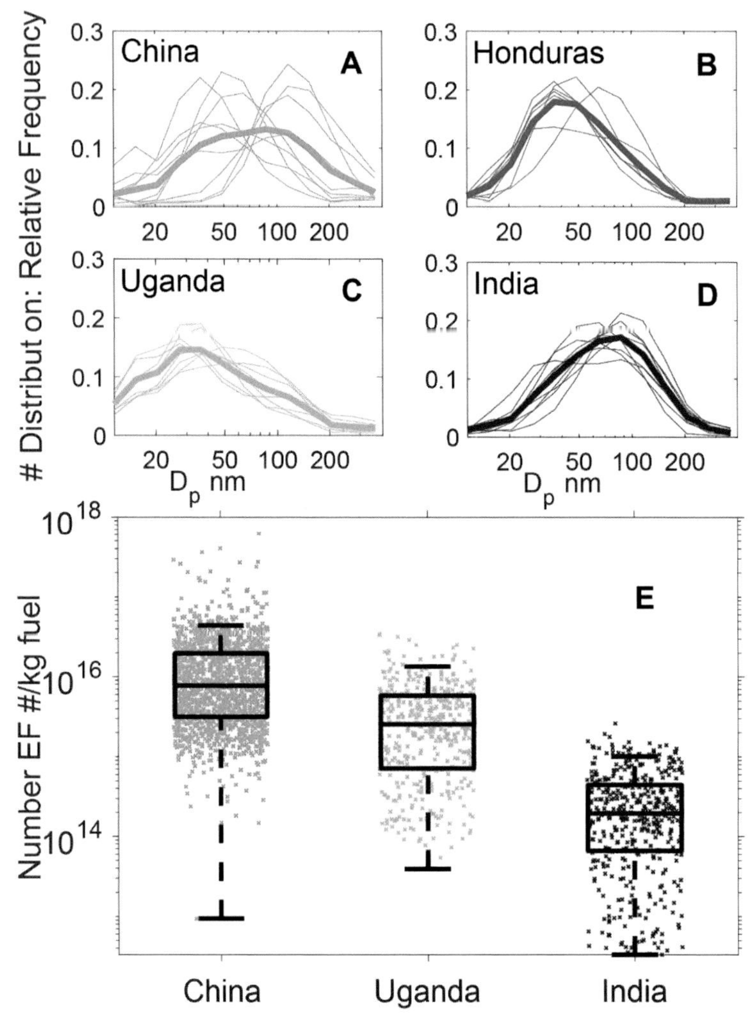

Fig. 8.2 Particle size distribution and particle number count in unconstrained cookstove emissions.[6] *(Courtesy copyright holder.)*

cookstove environment in these tests to be more toxic based on particle size and counts, despite the lower $PM_{2.5}$.

The Honduras experiment encountered technical difficulty, causing particle count not to be reported, making a comparison with India infeasible. In all cases of wood or charcoal combustion (Uganda, Honduras, and India), particle sizes are mostly in the ultrafine region. The authors of the report comment only on the impact of this on climate effect modeling. We see the significance to be in the health effects suffered by the cooks and others in proximity. We have shown in this book substantial evidence for the higher toxicity of ultrafine particles when compared with larger ones. This is especially the case when they carry an organic payload, as POAs do. This observation may also explain the high mortality in the cookstove operator group, roughly 70% of all PM-based mortality, which is disproportionate with the 20% overall atmospheric loading from unrestrained cookstoves.[7]

The immediate environment of the cookstove does not have confounding factors from volatile organic chemicals (VOCs) emanated from other sources. In fact, one would expect no secondary organic aerosols at all. The stove operators would inhale just the ECs and POAs. Accordingly, in vitro or in vivo studies of physiologic damage by these emissions could be expected to predict the health outcomes for stove operators. But in vitro studies could underestimate the impact because resuspension of filter-derived particles, which is required in such tests, causes agglomeration and loss of the physiologic functionality of the small size. Chapter 7 gives a fuller explanation of this phenomenon and remedies.

Engineered interventions

The earliest interventions have been improvements to existing cookstoves. Most LMICs have indigenous efforts at improving the emissions performance. Virtually all are directed at containment of emissions and improved efficiency of fuel combustion. Containment is achieved through the oven being more of an enclosure as opposed to an open front. Chimneys are added or made more effective. Laboratory comparisons have demonstrated reductions in PM emissions. However, field investigations have reported variable results.[5] In Fig. 8.1, the data in green are from improved versions. Although they trend lower than those for conventional versions for total $PM_{2.5}$, the differences are not statistically significant in most cases. Not surprisingly, the EC proportions are higher (see the India figures for the best clarity on this point). This would be expected due to the higher efficiency combustion leading to more complete burn and hence higher EC. Because EC particles are less toxic than POAs, the physiologic impact can be expected to be better, or less bad, for the emissions from improved stoves. More efficient fuel utilization will decidedly lower the PM associated with each period of use.

Another intervention is the use of different fuels. The commonest is charcoal that is wood slow-combusted under oxygen-starved conditions.

Volatiles are removed, especially water, leaving behind a concentrated carbonaceous material. This material will burn at a higher temperature than the original wood from which it was made, largely because of the water removal. It is less smoky as well, and so less associated PM could be expected. This is evident from Fig. 8.1, where the Uganda data are all from charcoal fuel, when compared with those of Honduras and India, both of which use wood. However, as already noted, the Uganda charcoal-based emissions are dominantly in the ultrafine range, with a high ratio of OC:EC, rendering them potentially more toxic on two counts. Also, the particle count is an order of magnitude greater than in the India observations (Honduras data not available). This one data set would appear to point to the use of charcoal in cookstoves, decreasing the mass of $PM_{2.5}$ emissions but likely increasing the health hazard, an interesting example of mass-based regulations driving behavior that may not offer any health benefit. More on this in Chapter 14, where we discuss policy implications of ultrafine particles.

Explicit comparisons of improved cookstoves were conducted by various investigators. Here we report selectively, choosing recent publications. Mitchell et al. studied three stoves in SubSaharan Africa, selected for significant differences in design (Fig. 8.3).[9] These were the Gyapa; a carboNZero rocket-type wood stove; and a radical departure, the Lucia, a pyrolysis or

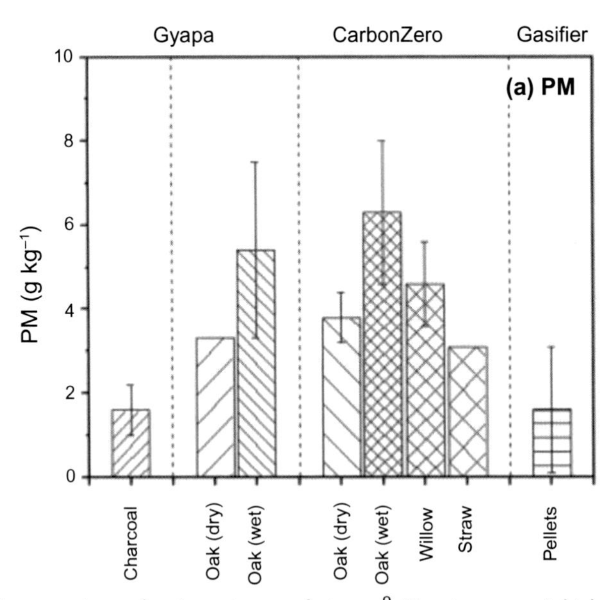

Fig. 8.3 PM comparisons for three types of stoves.[9] *(Courtesy copyright holder.)*

gasifier stove. The Gyapa had a raised hearth with perforations through which air was introduced from below and controlled by a damper; it was designed for charcoal but was commonly used with wood as well. The Lucia was a tube furnace in which controlled pyrolysis of wood was done; the output is flammable gases that were introduced into the burner in the stove; the residue was a char that may be used as an agricultural soil amendment.

The results are reported simply as "PM," but the size distribution data demonstrate all particles as being within the PM_1 classification, with most peaking in the ultrafine region. This behavior is generally similar to that reported in the Eilenberg study earlier. Raw wood, whether wet or dry, performs similarly in the Gyapa and carboNZero models. Charcoal PM emissions are much lower in the Gyapa, almost a third of those with wood. The Lucia gasifier performs the best in this parameter, with numbers similar to those of the Gyapa on charcoal feed.

The particle size distributions are sometimes bimodal, but all tend to have peaks close to 100 nm, with the singular exception of the gasifier, where the peak is close to 12 nm. Notably, although the gasifier and the Gyapa on charcoal have similar PM EF that were much lower than those of the other cases, the particle count is the highest for the Lucia. This is likely due to the small median size. This too was observed in the Eilenberg study for the Uganda charcoal data (Fig. 8.2). Another study of the gasifier type stove with lower PM EF showed 24 nm median diameters and a high proportion of counts under 35 nm. This is yet another demonstration of our observation that reduction in mass of PM does not necessarily result in fewer particles, and particle count, especially in the ultrafine sizes, is correlated with health outcomes.

The Mitchell et al. investigation also estimated the toxicity of the particles under the various conditions. The particles were collected and resuspended in cell culture media. The solution was pipetted into a six-well array containing cultured human lung epithelial cells, and single-cell gel electrophoresis was used to determine the DNA damage that was reported as the percentage of DNA tail per milligram of particle tested. Larger values indicate greater toxicity.

Fig. 8.4 plots toxicity (EF) as a function of particle size (expressed as count median diameter) denoted by the green circles and dashed line and particle count denoted by the pink squares and dashed line. Toxicity is weakly correlated with particle number EF ($R^2 = 0.5$, $n = 5$). Similarly, count median diameter is also weakly correlated with toxicity that is higher at the smaller diameters ($R^2 = 0.63$, $n = 5$). Elsewhere in the book, we show

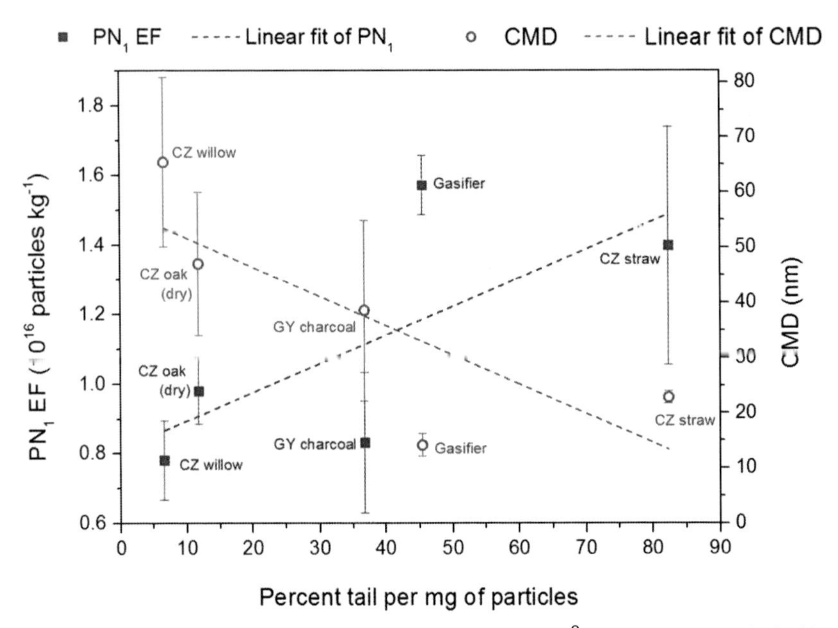

Fig. 8.4 PM toxicity as functions of particle count and size.[9] *(Courtesy copyright holder.)*

that physiologically one would expect the smaller particles, especially those under 100 nm, to be more toxic because of their greater ability to enter cells. But even more telling is that the resuspension method is undercounting the toxicity because ultrafine particles tend to agglomerate in the processing (see Chapter 7). One would expect, therefore, that the correlations would improve if the direct air–liquid interface method[10] were to be used to assess toxicity. In particular, the gasifier points in Fig. 8.4 could be expected to move to the right, to higher toxicity values.

An interesting take on cookstove emissions and their effect on human health is in a Sri Lanka-based study by Chartier et al.[11] They measured $PM_{2.5}$ during normal cooking operations on both standard and improved stoves, the improvements being in the containment and fuel efficiency. They demonstrated significant improvement in indoor $PM_{2.5}$ with the use of chimneys. The authors' hypothesis was that the actual PM loading experienced by the cook and companions was different from the loading in the room. To test this, the operator of the stove also wore a MicroPEM on a camera lanyard. This device is described in Chapter 6 and can capture any size fraction desired. The $PM_{2.5}$ estimated on this device is compared with the indoor levels in Fig. 8.5.

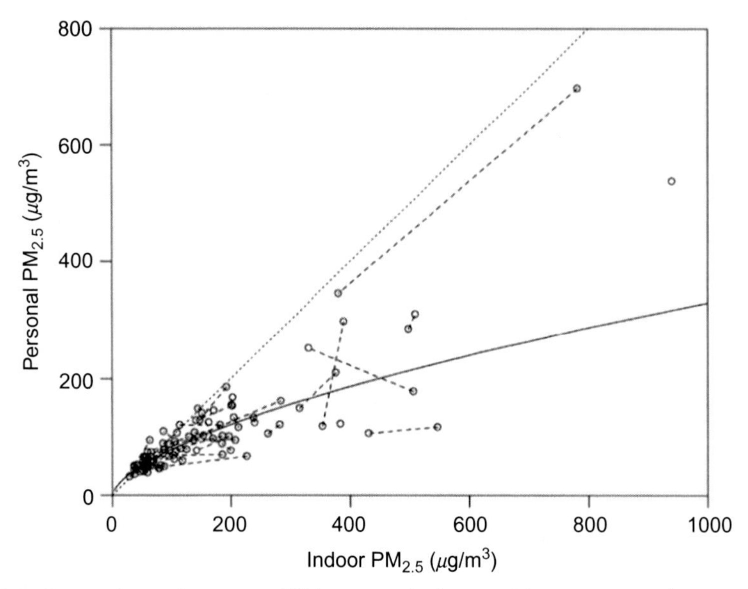

Fig. 8.5 Comparison of measured $PM_{2.5}$ on an indoor ambient sensor and a sensor on the person of the operator.[11] *(Courtesy copyright holder.)*

To assure the temporal coincidence of the compared data, they monitored compliance with the protocol by data from an onboard accelerometer that recorded the duration of the monitor being set aside (not worn while active). Compliance was close to 90%.

Fig. 8.5 demonstrates the nonlinear relationship between ambient and personal measurement. Ambient data overestimate operator exposure at the higher concentrations. This also means that improvements that manifest in either chimney or room measurements will not be reflected proportionately at the operator. But the overall finding is still that the operator suffers an insult lower than previously believed and that personal monitoring may have a place. Note also that the results are reported in mass per cubic meter (which is the metric of regulations) and not as EF.

The most dramatic interventions are the use of cookstoves fueled by kerosene, ethanol, methanol, or LPG that is predominantly a mixture of propane and butane. All these are liquids and can be effectively mixed with combustion air prior to ignition. This ensures a more efficient burn. The equipment is a radical departure from indigenous masonry or ceramic stoves but in principle not any more a departure than gasifier stoves such as Lucia. Various investigations have compared these with standard stoves. Discussed below is one of the most recent[12] findings. These authors reported using the

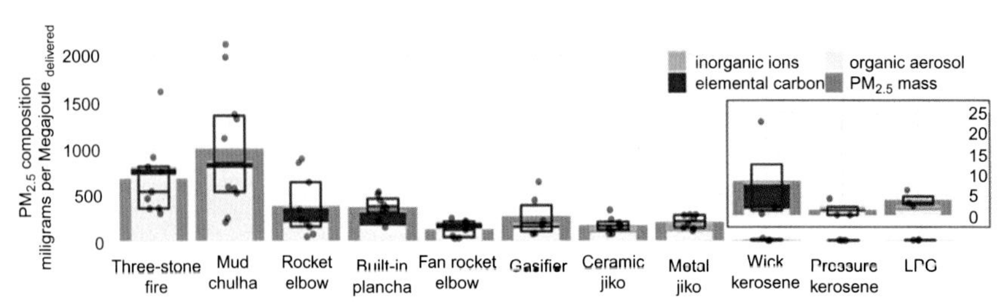

Fig. 8.6 Relative emissions of $PM_{2.5}$ from traditional and improved cookstoves, including hydrocarbon stoves.[12] *(Courtesy copyright holder.)*

parameter milligram per unit of energy produced, as opposed to per unit mass of fuel burnt. The PM figures are shown in Fig. 8.6. The three-stone stove is taken as the baseline. The kerosene and LPG stove performances are blown up to show detail. In keeping with previous reports, the improved wood stoves, such as the rocket elbow and the charcoal stoves, have improvements in PM of up to 80%. The kerosene and LPG stoves were over 99% lower on $PM_{2.5}$. This is to be expected, in part because of the mixing with air. In the case of the LPG, the small molecules comprising the propane and butane could be expected to produce minimal soot. The organic aerosol portion of the PM showed a greater increase than did the EC. Although the epidemiologic proof is still lacking, mechanistic considerations (relative toxicity of organics versus carbon) suggest that OC is more toxic than EC.

Confirming previous observations, the gasifier stove, although lower on PM count, produced more ultrafine particles than the three-stone base case, as did the fan rocket elbow. All the other improved stoves were lower on ultrafines as well. The best performers were the two kerosene versions, at over 95% reduction, and the LPG at 89%. However, the Wick kerosene stove produced significantly greater VOCs such as benzene and formaldehyde. The balancing act of reduced $PM_{2.5}$, ultrafines, and VOCs likely needs further definitive studies on health effects to select winners among the improved stoves. This is particularly the case because the economics are different, both on equipment cost and the variable cost of the fuel. However, at first blush, discouragement of kerosene is probably merited in favor of LPG. Countries with heavy kerosene subsidies, such as India, may be reluctant to make the switch but that is for discussion in a later chapter.

Ethanol-based cookstoves were reported by Benka-Coker et al. for the implementation of CleanCook stoves in refugee camps in Ethiopia.[13] They reviewed the work of others with that stove, indicating 99% reduction in $PM_{2.5}$ when compared with the standard three-stone fireplace. Based on chemistry (no chain of carbon-to-carbon bonding in the molecule), one would expect this sort of performance on reduction in PM. One could also expect methanol-fueled stoves to have similar performance.

Economics of substitutes

Yamamoto et al. did a study in Burkina Faso, a landlocked country in West Africa. Based on per capita income, it is one of the poorest countries in Africa. Ninety five percent of the population used biomass for cooking. The mortality ascribed to the indoor biomass smoke from cooking was 21,500 annually, in a population of just 14.4 million.[14]

The study discussed fuel substitutes and the associated economics. Lacking Burkina Faso specific data on fuel consumption, they used data from India as a proxy. In India, roughly 1 MJ of energy is consumed per meal. The authors estimated that in Burkina Faso, the fuel consumption would be 1.6 MJ per day. Using local pricing of commodities, they found that the cost for using wood would range from USD 1.64 to 2.45 per year and that charcoal would cost USD 4.25. LPG was estimated to cost USD 44.3, a high fraction of the average annual income of USD 162. They nevertheless concluded that the benefits are worth the cost, presumably subsidized by the government. Here, we offer a potentially different solution: domestically produced methanol.

Most of the countries with a prevalence of biomass fuel usage are net importers of hydrocarbons. LPG would be an imported commodity. On the other hand, many of the countries have an abundance of coal, especially China and India. Coal can be converted to methanol at a cost under USD 300 per tonne (in China that figure is USD 250). The close-up below describes methanol synthesis and the associated economics.

A kilogram of methanol has an energy content of 20 MJ. At 1.6 MJ per day, the annual requirement for a household would be 584 MJ. This translates into 292.2 kg of methanol. Using the USD 250 per tonne figure, the cost would be USD 7.25 per year. This is much more affordable than LPG and amenable to a relatively small subsidy to get it into charcoal territory. In India, kerosene is heavily subsidized for use as a cooking and lighting fuel. A transfer of the subsidy to methanol would be a simple policy change. Parity with wood is not necessary because wood gathering has additional social and economic costs.[14] Collecting wood can take up to 4.7 h per week. Women performing this function may attempt to reduce this time by collecting less, substituting with inferior biomass, or simply cooking fewer meals. Other considerations are loss of income, hygiene, and personal safety.

Close-up: The case for methanol as a cookstove fuel

The best performing fuel substitutes for biomass appear to be ethanol and methanol. The performance measures are $PM_{2.5}$ mass per unit of energy produced and ultrafine particle count. LPG will come close but will always be more expensive. The entire focus of this discussion is the developing world (LMICs).

Ethanol is most economically produced from fermentation of sugars. These sugars are derived in the main from sugarcane, sugar beet, and sweet potato. Corn is also a significant source. Consumption of any of these to produce ethanol competes with food. Also, in many of the LMIC areas, if these are grown, they are considered cash crops, generating necessary income and likely foreign exchange.

Methanol, on the other hand, can be synthesized from a variety of raw materials, some of which are wastes. Any biomass can be gasified to produce synthesis gas, known as syngas, from which gasoline, diesel, dimethyl ether, or methanol can be made. Of these, methanol synthesis is the simplest from the standpoint of process complexity and cost. Coal of any grade can also be gasified. Most commercial methanol is synthesized from natural gas, again with syngas as an intermediate step. In LMICs, the most likely source of methane is biogas. In a village setting, this could be from animal wastes, a cleaner alternative to using the waste directly as a fuel, as in the case of dried dung. In some areas, such as the state of Assam in India and portions of West Africa, oil production results in the production of associated natural gas, which is currently flared. The conversion of such gas to a liquid has been hampered by the economics of transport to major plant locations. Technologies to perform this economically on a small scale are being researched, and one such is already near commercial.[15] In many cases, animal wastes may also be in small enough quantity to also require such a small-scale reactor.

The knock on methanol has been that it is poisonous if ingested in the belief that it is ethanol. Each country would need to arrive at measures most conducive to societal norms to prevent this. One measure already used is the addition of a bitterant. Another might be to give it a brand name and not be transparent about the ingredient being an alcohol.

Two large users of biomass cookstoves are China and India. Both, and some other smaller players as well, are net importers of oil and gas, from which kerosene and LPG are derived. The price is proportional to the world price of oil and has fluctuated over the years. This makes kerosene and LPG unreliable fuels for use by the poor. Domestic coal, however, is plentiful, although much of it is of low grade (high moisture and/or ash). In our opinion, conventional gasifiers are not well suited to high-ash coal. Low-grade coal such as lignite may have ash content of up to 30%, when compared with single digits for the higher grade bituminous coals.[16] Conventional gasifiers operate above the fusion temperature of the fluxed ashes (they are in the molten state and removed in that state). At high ash content, this challenges the linings of the reactor.

The U-GAS process is a fluidized bed reactor operating at temperatures below the fusion point of the ash. It was devised by the Gas Technology Institute (GTI), with commercial offerings exclusively by Synthesis Energy Systems, Inc. Whereas it has been piloted in several countries, including India, significant size plants exist only in China, where the process has been perfected for the available coal, ranging from bituminous to lignite. Coal or biomass is fed through a chute into the lower area of the reactor. Air or air enriched with oxygen is fed in from below, together with steam that is essential for an aspect of the reaction. The ash is removed from the bottom of the reactor in solid form, sometimes agglomerated, as shown in Fig. 8.7.

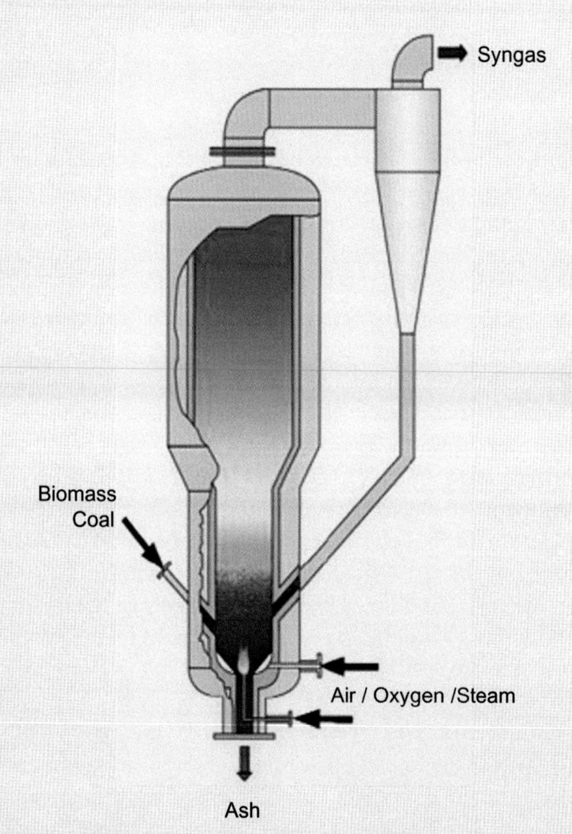

Fig. 8.7 Schematic of the U-GAS process for gasification of coal and biomass. *(Courtesy: Gas Technology Institute.)*

The output is syngas with a prescribed ratio of hydrogen to carbon monoxide. For methanol synthesis, the ratio will be somewhat under 2.0. The nature of the process equipment allows the plant to be relatively small. Conventional methanol plants are sized between 5000 and 10,000 tonnes per day. The U-GAS plants are as small as 700 tonnes per day.[17] This feature could be useful, where the target market is village settings. The reported cost of production in China is USD 250 per tonne of methanol. That is the figure we used in the body of the chapter to estimate the cost to users of cookstoves fueled by methanol.

When the available raw material for methanol synthesis is in relatively small supply, although proximal to the end use point, even the U-GAS process may not be feasible. The MicroReformer is a small-scale syngas generator conceived by the Massachusetts Institute of Technology (MIT) and scaled up by RTI International. It overcomes the economies of scale hurdle by using a mass-produced reactor, a modified diesel engine.[15] The feed for this reactor is methane from any source, including natural gas that would otherwise be flared and biogas from animal waste. These sources can be as small as 50,000–200,000 standard cubic feet per day. The MicroReformer can operate profitably at these volumes, producing 2–10 tonnes methanol per day. Flared gas will likely have some component of larger molecules than methane, but the process is able to handle those. Depending on the value ascribed to the input gas, the delivered cost of methanol will be somewhat different but still in the vicinity of that produced in conventional plants.

The market price for methanol also is proportional to the price of oil. At the time of this writing, the price in Asia is USD 245 per tonne and has been double that not too many years ago. Consequently, it too is an unreliable substitute for biomass in cookstoves. It would have to rely on government subsidies in the form of price control or direct subsidies, as is the case with kerosene today in many countries. The most viable strategy is dedicated production of methanol for cookstove use. This could be through public sector undertakings or price control mechanisms such as utility commission-managed electricity supply in the United States. This last came about because continuous electricity supply was believed to be a right. Perhaps methanol ought to be viewed in the same light for cookstove users.

Large-scale coal to methanol plants should likely be the backbone of the supply. Where waste gas or biogas pockets are available, the small-scale converters may make the most sense. In some respects, this combination would look like large electricity generators with grid supply to villages, augmented by local microgrids supplied by solar or wind. To the extent that the methanol must be subsidized, the purely economic justification for it is reduction in health costs and the recovery of the cost of loss of productive human life.

References

1. Schwarzenbach RP, Gschwend PM, Imboden DM. *Environmental organic chemistry*. 2nd ed. Hoboken, NJ: John Wiley & Sons; 2005 [p 298].
2. Curkeet R. Wood Combustion Basics. In: *Presented at EPA Workshop March 2, 2011*; 2011.
3. World Health Organization, May 8th 2018. Household Air Pollution and Health Fact Sheet. http://www.who.int/en/news-room/fact-sheets/detail/household-air-pollution-and-health, [Retrieved April 7, 2020].
4. Sehgal M, Garg A, Goel A, Mohan P, van den Hombergh H. *Cooking fuels in India: Ttrends and patterns*. Delhi: The Energy and Resources Institute; 2010. https://www.teriin.org/sites/default/files/2018-05/policy-brief_cooking-fuels-in-India.pdf. [Retrieved April 14, 2020].
5. Roden CA, Bond TC, Conway S, Pinel AB, MacCarty N, Still D. Laboratory and field investigations of particulate and carbon monoxide emissions from traditional and improved. *Atmos Environ*. 2009;43(6):1170–1181. https://doi.org/10.1016/j.atmosenv.2008.05.041.
6. Eilenberg RS, Bilsback KR, Johnson M, et al. Field measurements of solid-fuel cookstove emissions from uncontrolled cooking in China, Honduras, Uganda, and India. *Atmos Environ*. 2018;190:116–125. https://doi.org/10.1016/j.atmosenv.2018.06.041.
7. Bond TC, Streets DG, Yarber KF, Nelson SM, Woo JH, Klimont Z. A technology-based global inventory of black and organic carbon emissions from combustion. *J Geophys Res: Atmos*. 2004;109(14):1–43.
8. Lacey F, Henze D. Global climate impacts of country-level primary carbonaceous aerosol from solid-fuel cookstove emissions. *Environ Res Lett*. 2015;10.
9. Mitchell EJS, Ting Y, Allan J, et al. Pollutant emissions from improved Cookstoves of the type used in sub-Saharan Africa. *Combust Sci Technol*. 2019. https://doi.org/10.1080/00102202.2019.1614922.
10. Lichtveld K, et al. In vitro exposures in diesel exhaust atmospheres: resuspension of PM from filters verses direct deposition of PM from air. *Environ Sci Technol*. 2012;46(16):9062–9070.
11. Chartier R, Phillips M, Mosquin P, et al. A comparative study of human exposures to household air pollution from commonly used cookstoves in Sri Lanka. *Indoor Air*. 2017;27:147–159. https://doi.org/10.1111/ina.12281.
12. Bilsback KR, Dahlke J, Fedak KM, et al. A laboratory assessment of 120 air pollutant emissions from biomass and fossil fuel Cookstoves. *Environ Sci Technol*. 2019;53(12):7114–7125. https://doi.org/10.1021/acs.est.8b07019.
13. Benka-Coker ML, Tadele W, Milano A, Getaneh D, Stokes H. A case study of the ethanol CleanCook stove intervention and potential scale-up in Ethiopia. *Energy Sustain Dev*. 2018;46:53–64. https://doi.org/10.1016/j.esd.2018.06.009.
14. Yamamoto S, Sié A, Sauerborn R. Cooking fuels and the push for cleaner alternatives: a case study from Burkina Faso. *Glob Health Action*. 2009;2. https://doi.org/10.3402/gha.v2i0.2088.
15. Rao V, Knight R. *Sustainable shale oil and gas: analytical, geochemical and biochemical methods*. Boston: Elsevier; 2017:201.
16. Bowen BH, Irwin MW. *Coal Characteristics*; 2008. https://purdue.edu/discoverypark/energy/assets/pdfs/cctr/outreach/Basics8-CoalCharacteristics-Oct08.pdf. [Retrieved April 14, 2020].
17. Lau F. *Commercial Development of the SES UGAS Gasification Technology*; 2009. https://www.globalsyngas.org/uploads/eventLibrary/32LAU.pdf. [Retrieved April 14, 2020].

CHAPTER NINE

Wood fires: Wild

Wood fires in cookstoves are a necessity for 3 billion people. The particulate matter (PM) emitted from these fires is responsible for up to 4 million premature deaths per annum and is the single biggest public health problem in the world.[1] In comparison, involuntary wildfires appear to pale in significance but they merit examination on several grounds. Due to several factors, including climate change-induced increase in droughts and rising temperatures, poor forest management, and a growing population, wildfires are projected to increase in frequency and intensity over the next century.[2] This increase, aside from causing privation for people in proximity, will inevitably lead to increases in ambient PM that is at least as toxic as that from cookstoves. This effect will be felt far from the origins of the fires. In this chapter, we discuss the causes of the fires, the health effects due to emissions, and possible means to slow down the upward trajectory of incidence.

Wildfire emissions

The fundamentals governing wildfires are the same as for controlled fires in cookstoves. Wood species and combustion conditions (flaming versus the other extreme, smoldering) are the key determinants of the size and composition of the particles emitted. But there are differences that matter. Domestic fires are from wood chosen based on proximity and burn characteristics. Wildfires are not choosy. For example, pine is not a preferred cooking wood but can represent the dominant species in some fire areas. The same fire can have differences in combustion conditions and resultant particle character within hours. An example is the so-called Getty Fire in southern California. It was accidental and close to housing. The main flames were doused quickly, but within a few hours flaming and smoldering occurred. For each of the periods marked by Roman numerals in Fig. 9.1, the particle size distribution is significantly different, with the peaks moving by many tens of nanometers.[3] This underlines another variable in wildfires: the duration of unconstrained burn versus the smoldering phase.

Particulates Matter
https://doi.org/10.1016/B978-0-12-816904-9.00018-0

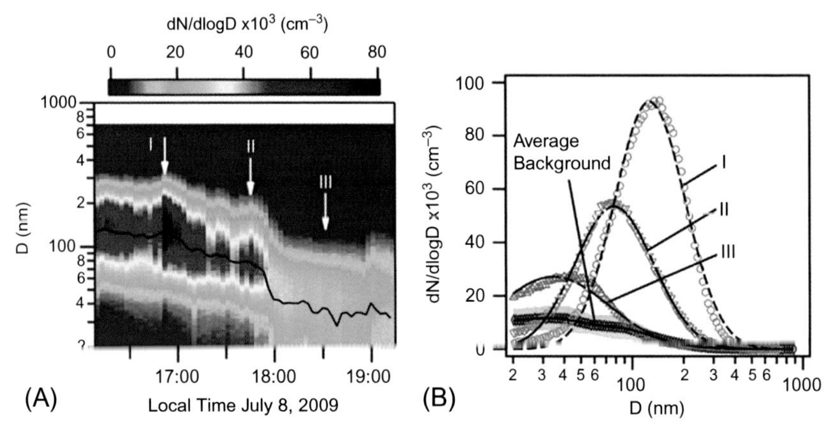

Fig. 9.1 Time-distributed particle size ($D =$ diameter), with the log normal fit median diameter shown as the *solid line*. *Color* indicates concentration. A log normal size distribution is described by d*N*/dLog *D*, where *N* is the particle count; the peak of the distribution is shown at the median diameter. The second panel shows particle count as a function of diameter for each zone marked by *arrows* in the first panel, with the log normal fits shown as *dashed lines*.[3]

In principle, one could expect there always to be these two conditions, just different durations.

The particles are distributed in and near the ultrafine range but are larger than those from cookstoves (Chapter 8). From this, one could tentatively conclude that the toxicity would be less. As discussed later in this chapter, that is generally not the case. Another feature of wildfires is that, in addition to the elemental carbon and organic carbon particles, there is also ash. In cookstoves, the ash is retained in the stoves and collected. In wildfires, especially in the windy conditions with which they are commonly associated, the ash is airborne. Ash is inorganic, being the noncombustible portion of the biomass, and ought to be less toxic.

The median particle diameter dropped from 130 nm in phase I to 40 nm in phase III. The authors of the study considered several possibilities for this rapid decrease, including moisture and photochemical action. They concluded tentatively that the reason was the shift from flaming to smoldering, which occurred quickly. This fits their laboratory studies, where flaming to smoldering resulted in a decrease from medians of 40 nm to 20 nm. These numbers are much lower than the real-life numbers and more like the cookstove numbers. Our takeaway is that the mechanisms are similar, but the particles are generally larger than in laboratory studies.

These results are for a specific fire. The authors cite studies of other fires in the same season with larger median particle diameters and particle size distributions peaking at up to 250 nm. Another investigator studied smoke from the wildfire-prone species pine, eucalyptus, and oak and found single mode peaks in the 100–200 nm range.[4] A US Environmental Protection Agency advisory notes that wildfire emissions have PM in the visible range of wavelengths, 400–700 nm, causing scatter of incident light and making for poor visibility. Whereas size has significance for health effects, it also affects transport, in that smaller particles remain suspended and travel further downwind than do larger ones.

General health effects

$PM_{2.5}$ (particles with diameter less than 2.5 μm) from wildfires are reported to have caused an average 339,000 premature deaths annually worldwide during the period 1997–2006 (Table 9.1).[5] The bulk of these, unsurprisingly, are in low- and middle-income countries. While the number is small when compared with deaths associated with domesticated fires (cookstoves), they are still large and warrant study for amelioration. There were significant difference between the averages in the (hot) El Niño and (cold) La Niña years. Southeast Asia appears to be the most sensitive; El Niño there is associated with drought, a contributor to wildfires. These data add to the body of evidence connecting climate change with wildfires and the associated mortality. The only caveat is that the 1997–98 El Niño was one of the

Table 9.1 Mortality ascribed to wildfires in El Niño and La Niña years.[5]

Scenario	Global	Sub-Saharan Africa[a]	Southeast Asia[b]	South America[c]
Annual average (1997–2006)	339,000	157,000	110,000	10,000
EL Niño year (September 1997–August 1998)	532,000	137,000	296,000	19,000
La Niña year (September 1999–August 2000)	262,000	157,000	43,000	11,000

[a]WHO subregions 18–21.
[b]WHO subregion 5 only.
[c]WHO subregions 11–14.
Results are shown for the three most severely smoke-affected regions. These estimates are based on the assumptions used in the principal analysis.
Data from Johnston F, Hanigan I, Henderson S, Morgan G, Bowman D. Extreme air pollution events from bushfires and dust storms and their association with mortality in Sydney, Australia 1994–2007. *Environ Res*. 2011;111(6):811–816. https://doi.org/10.1016/j.envres.2011.05.007.

most intense on record. Also, the anomalous data in sub-Saharan Africa could be attributed to the fact that the drought there was surprisingly mild. Kenya had an unusually wet season, as it usually does in El Niño years (see Fig. 9.7; the dark green oval depicting wet conditions covers Uganda and parts of Kenya and Tanzania). The inescapable point is that mortality appears to be tied to relatively short-term effects.

Another study[6] projected mortality from $PM_{2.5}$ emissions in the continental United States for the years 2050 and 2100, with 2000 as baseline. The equation was

$$\Delta\,\text{Mortality} = \text{Pop}\left(1 - e^{\beta \times \Delta X}\right) Y_0, \tag{9.1}$$

where Pop is the population, Y_0 is the baseline mortality, ΔX is the change in the annual mean $PM_{2.5}$ concentration, and the β coefficient is the concentration-response factor derived from the relative risk determined from many epidemiological studies. Johnston et al. in the study cited above likely used similar methods to arrive at the β coefficient because the results are comparable. Fig. 9.2 shows the Ford et al. results, broken down by greenhouse gas source: nonfire, fire, and atmospheric transport (e.g., by wind). For the two projected time points, estimates are made for two representative concentration pathways (RCPs), 4.5 and 8.5 W/m^2 (Watts per square meter). RCP is defined as the radiative forcing in the year 2100 (for a discussion of radiative forcing see Chapter 4). The pathways are computed with models using assumptions that are generally accepted by the scientific

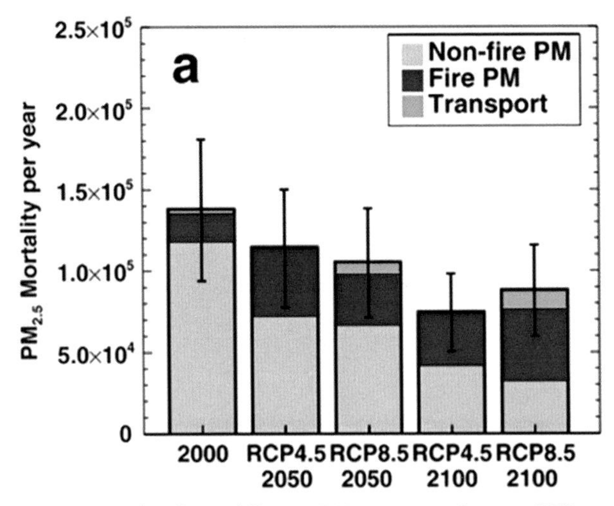

Fig. 9.2 Estimated mortality from different $PM_{2.5}$ sources for two RCP scenarios.

community. The RCP 8.5 level is not accepted as realistic by a few scientists,[7] but the general belief that not enough is being done to curb greenhouse gas emissions leads to considerable support for using 8.5.

Fig. 9.2 suggests that mortality from $PM_{2.5}$ from nonfire sources will decline over time, indicating progress in reducing some anthropogenic emissions. But wildfires increase their relative contribution in both scenarios in 2050 and 2100. The authors did not estimate the atmospheric transport component for RCP 4.5. This component accounts for PM in transported smoke from Alaska, Canada, Hawaii, and Mexico. Their model suggests that in the RCP 8.5 scenario, wildfire-related mortality will become the dominant component in 2100.

PM from wildfires has been shown to cause lung toxicity, with several studies demonstrating greater effects than PM from other sources. An in vivo study using mice collected wildfire-derived PM and administered it through the trachea, rather than through the breathing process. The investigators used a bolus dose in the physiologic range, about what would be expected if the mouse had inhaled it. They found that alveolar macrophage damage was the principal toxic effect. The impact was greater than that of the same dosage of particles from ambient air.[8] The finding that bolus PM_{10} and $PM_{2.5}$ had the same damaging effect is likely an artifact of the intratracheal delivery. Inhalation delivery would have captured some of the coarse particles in the nasal cavity. But the physiologic mechanisms should remain the same, especially for $PM_{2.5}$ that importantly also contains the ultrafine portion.

Lung toxicity

The lung is the principal organ of interest because inhalation is the most common means for wildfire PM to make its way into the human body. Impact on that organ may be estimated by in vitro and in vivo methods. In vitro techniques primarily involve collecting size fractions of interest on filters, resuspending them in a culture, and administering the resuspension to lung epithelial cells. Toxicity is estimated by examination of the cells. An improvement using direct air-liquid interface mechanisms is more physiologically relevant but relatively new. In Chapter 7, it is described and compared with the resuspension method.

In vivo techniques use mice or rats. The animals are placed in a chamber with the ambience loaded with captured or simulated wildfire emissions. This is the most physiologically appropriate method. However, many investigators choose to introduce the emission-laden air, or a bolus of PM derived from it, intratracheally into the lung, which ensures uniform dosage. The science community appears to have accepted this method if the dose delivered in the bolus is at a physiologic level (as opposed to a pharmacologic level).

A recent study by Kim et al. is described in some detail because it used biomass types that are good proxies for forests in several countries and because it systematically investigated flaming and smoldering.[9] The definition for the combustion condition was modified combustion efficiency (MCE).

$$\mathrm{MCE}\% = \Delta CO_2/(\Delta CO_2 + \Delta CO), \tag{9.2}$$

where Δ represents the excess concentrations of each of the species. MCE describes the state of oxidation of the carbon, and 100% is essentially complete combustion to CO_2. Smoldering is defined as MCE 65%–85% and flaming as >95%.[10] The biomass types were red oak, peat, pine, pine needles, and eucalyptus. These are representative of the vegetation likely to be components of wildfires in the United States and are reasonably representative for some other countries, including Australia. Except for peat, all the fires demonstrate a linear relationship between PM emission factor (EF; for a definition see Chapter 8) and MCE (Fig. 9.3). The data from this study

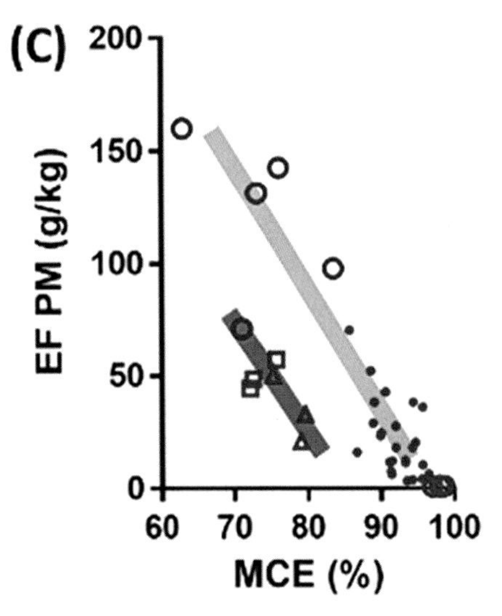

Fig. 9.3 Emission factors for particulate matter as a function of modified combustion efficiency.

are in open circles, the other data from other studies. The blue line is for above-ground biomass and the gray line is for on- or in-ground biomass. Unsurprisingly, the EFs are the highest in the oxygen-starved conditions of lower MCE. Not shown are their EF CO and EF CO_2 values that follow expected patterns from the definition of MCE.

The lung toxicity measures were two markers of inflammation, neutrophils and macrophages (see the text box below). Also estimated were markers of cell damage. For lung toxicity EFs (toxicity per unit mass of fuel combusted), neutrophil values were used and expressed as number per kilogram of fuel combusted. With this metric, as opposed to number per mass of PM, smoldering conditions produced the most effect. (With the other metric, flaming producing higher values, a point of note when comparing different studies.)

Fig. 9.4 compares lung toxicity potencies of emissions from the five biomass types. The potencies were assessed from the number of neutrophils in bronchoalveolar lavage fluid based upon equal mass of PM. Mice were exposed to the PM through oropharyngeal aspiration. This technique introduces the substance into the pharynx that is located at the back of the mouth and is subsequently aspirated into the lungs. The method is less invasive than intratracheal delivery. Neutrophils were counted at 4 h and 24 h after exposure. Lung toxicity is the highest for eucalyptus at both time points and under all combustion conditions. The toxicity difference between smoldering and flaming is striking when the metric is neutrophils per unit mass of fuel, which is a reasonable metric in assessing the impact of forest fires.

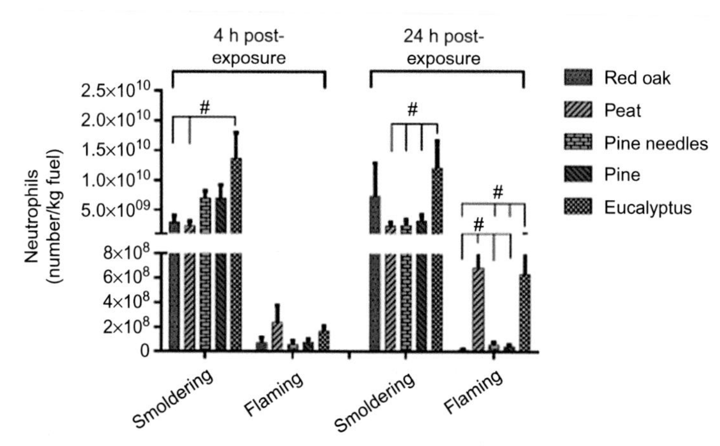

Fig. 9.4 Lung toxicity measured as neutrophils per unit of bronchoalveolar lavage fluid for different biomass and burn conditions at 4 h and 24 h after exposure.

The general point to be taken is that toxicity is a function of biomass type and combustion conditions. While none of this is surprising, the biomass type identification is new. Not good news for the Australians (who have an abundance of eucalyptus, the most toxic biomass in the study). But Australians can take some comfort from the investigation of mutagenic potency. Here, the authors report that the worst actors are peat and pine, and each potency was much higher than that of other species, including eucalyptus.

Measures of health impact of particulate matter

The impact of PM may be assessed for the short and the long term. For the short term, the immune response is of primary interest. For the long term, it is mutagenicity leading to lung cancer.

The neutrophil is a type of white blood cell, the most abundant type. Kim et al. used high neutrophil count as a marker for inflammation and mutagenicity.

The macrophage (Greek for big eater) is another type of white blood cell. It patrols body tissues and engulfs foreign pathogens, in the process of phagocytosis. High macrophage count is also a marker of inflammation and infection.

Mutagenicity is the propensity for particles to cause mutation, altering the genetic material of an organism (the DNA), which can lead to cancer. Kim et al. used a *Salmonella* plate-incorporation mutagenicity assay, the description of which is beyond the scope of this author and this book. Suffice to say, they estimate the reproducible dose response.

Causes of wildfires

Wildfires have increased in severity and frequency in the last two decades.[11] This has prompted a harder look at intervention that goes beyond the traditional means of fire suppression and fuels management, such as controlled burns. Causes vary by country and regions within countries, but a common pattern is emerging that the majority are anthropogenic. The gamut runs from arson to the more indirect human-abetted climate change. We will discuss causes in France and California as reasonably representative, and even these have commonalities.

Ninety five percent of fires in California and the Mediterranean region are caused by humans. The fine statistics do vary. Ignition frequencies are instructive, but fire severity is not directly correlated. Fig. 9.5 shows proportional causes in two areas of southern California. Although the numbers

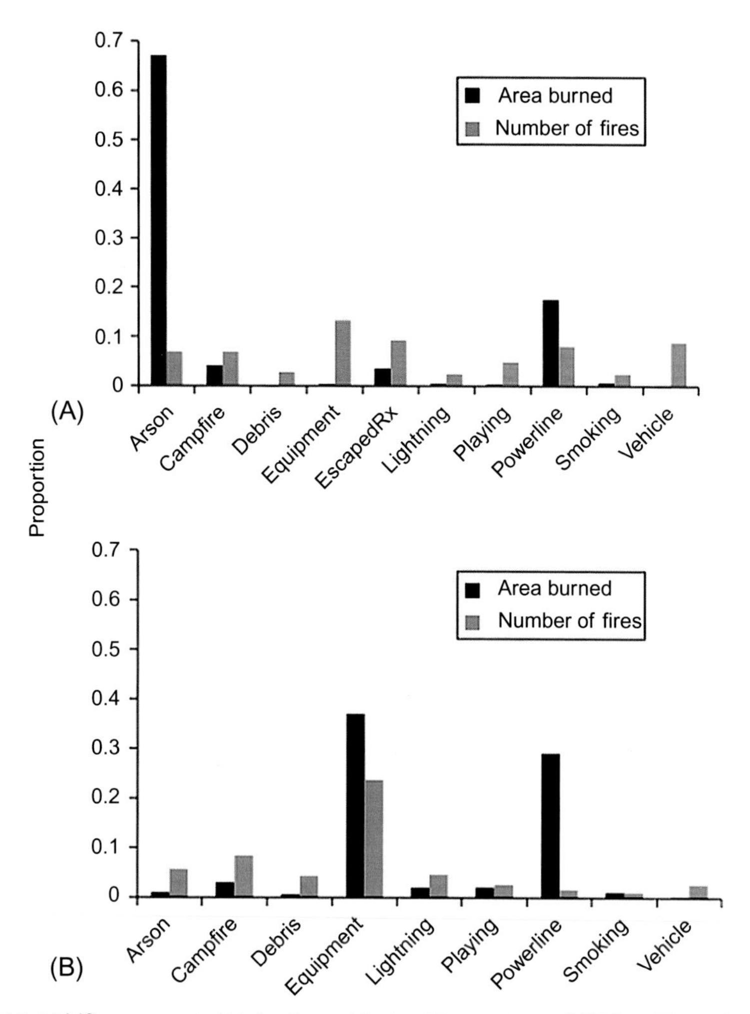

Fig. 9.5 Wildfire causes in (A) the Santa Monica Mountains and (B) San Diego County. "EscapedRX" stands for controlled burns in which the control was lost.[12]

ascribed to arson were close in the two locations, the area burned was vastly different. Of particular note are the powerline-initiated numbers. At the time of writing, this cause has essentially bankrupted the affected California power utility. The number of ignitions is small in San Diego County, and yet the area burned is disproportionately high. The natural ignition source lightning is low in number and severity in both areas.

The distance to a road was a contributor in roughly 40% of the cases in both areas (data not shown). This surprising statistic is supported by the fact that so many instances of human-induced ignition, such as arson, campfires, equipment operation, and smoking, are facilitated by road access. The category of equipment covers items such as sparks from mowers, power saws, and other machinery. Most ignition occurs at ground level, with grass often being the tinder.

Fig. 9.6 underlines the seasonal variability in the different ignition origins, expressed in the burn areas rather than ignition incidences. Power line and equipment incidences were dominant not only in October in both areas but also in July for power lines. The Santa Ana winds blowing in the fall are largely contributory to power line-caused ignition in those months. Interestingly, arson is also a late summer and fall event.

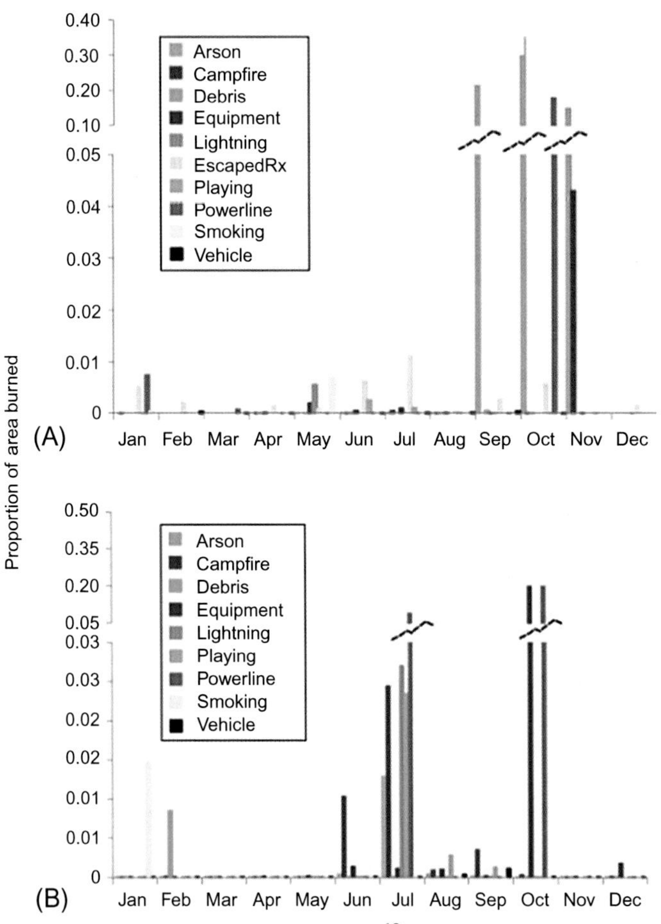

Fig. 9.6 Seasonal variation in wildfire burn area.[12]

Studies in southern France confirm the California results in many respects. Here, the two largest ignition causes were arson and equipment, and the incidents were concentrated in the summer. Severity was reduced by rain in the spring. The effects were regionally specific but had one character as a whole: 1% of the fires created 70% of the burn area.[13] The two studies have a common takeaway: ignition source and severity are highly regional even within parts of a country or state. But one ignition source, power lines, merits further discussion in part because, at the time of this writing, major California fires were laid at its feet, and the utility had admitted guilt for 84 deaths in the Camp Fire occurrence in Paradise.[14]

The role of power lines

2017 and 2018 were extreme drought years in California. Historically severe numbers of wildfire occurred, resulting in home destruction and directly related deaths (as opposed to deaths later due to resulting PM). In the winter of 2017 and again in the summer of 2018, a dozen or more fires were blamed on power lines. This can occur when live lines (generally not insulated) are impacted by trees blown down by wind, abetted by dry ground. Utility poles snapping in high winds and the fallen live wires igniting dry grass and brush is another mechanism. The Pacific Gas and Electric Company (PG&E) was blamed for the 12 fires in late 2017 and the 17 fires in the summer of 2018. The utility is in serious financial trouble and has taken to cutting power to communities during weather events considered likely to cause power line-mediated ignition.

Two main remedies were discussed by the community. One is buried power lines. Residential communities across the country have this feature, especially in new developments where trenching prior to building is less costly and the lines are not high voltage. The San Diego utility already has portions of its grid underground. That has prompted a call for this solution elsewhere. A *San Francisco Chronicle* article[15] estimates per mile costs in various jurisdictions where the utility operates. For the state as a whole, with much coverage in forests, the high voltage lines are expected to cost USD 5.0 million per mile, with a total bill for the state at USD 100 billion. Regulations require that the cost be borne by the ratepayers. This remedy is unlikely to be enacted.

The other approach is to insulate the currently bare wires so that when they contact trees or brush, no ignition occurs. Because these are high voltage lines of large diameter, and two problems arise. One is that the sheer weight increase due to the insulation will likely require the poles to be strengthened and possibly moved closer, adding cost. The other, a bit more insidious, is that the current flow in the wires generates heat that, with insulation, would not be dissipated as easily. Making the wires thicker would help, as would more wires, adding even more cost.

The role of climate change

The term "climate change" has many proxies. The most common is a rise in ambient atmospheric and ocean temperatures and the changes resulting therefrom. Rise in ocean temperature is connected to a phenomenon such as El Niño that in turn is responsible for catastrophic droughts and precipitation (in the same year but in different parts of the world). Rising temperatures can reduce the proportion of precipitation as snow. The amount of snow and timing of the snowmelt can affect the conditions favoring forest fires.

A prevailing viewpoint is that climate change causes more wildfires. Not uncommon are statements such as this from Anthony Westerling: "The timing, extent and severity of wildfire in western US forests is strongly influenced by climate."[16] But in fairness, the scientific community is divided on causality. The naturally occurring phenomenon of El Niño may be exacerbated by global temperature rise. In any case, even naturally occurring El Niño events will promote the conditions leading to wildfires.

El Niño and La Niña

El Niño is a naturally occurring phenomenon in the Pacific Ocean every 2–7 years. The frequency and severity appear to be on the rise, and some authorities connect it to global warming. It originates in the equatorial region, but its effects are felt far from the equator. In normal (non-El Niño years) years, trade winds blowing from east to west cause the warm oceanic waters to flow toward the west Pacific. This departure of surface warm water causes cold water and nutrients from deep locations to rise on the coasts of the Americas, reducing the temperature there and nourishing the organisms. In El Niño years, the trade winds weaken to the point that the warm waters stall and reverse toward the Americas. Now that the area is warmed instead of cooled, causing a reversal of the normal climatic conditions and usually generating rain and storms. On the other side of the Pacific, instead of warm and wet weather, they get droughts (Fig. 9.7).

The year following an El Niño often sees a reversal to an extreme case of the norm known as La Niña. The trade winds blow east to west but much stronger than normal. This results in heavy rain in the year following an

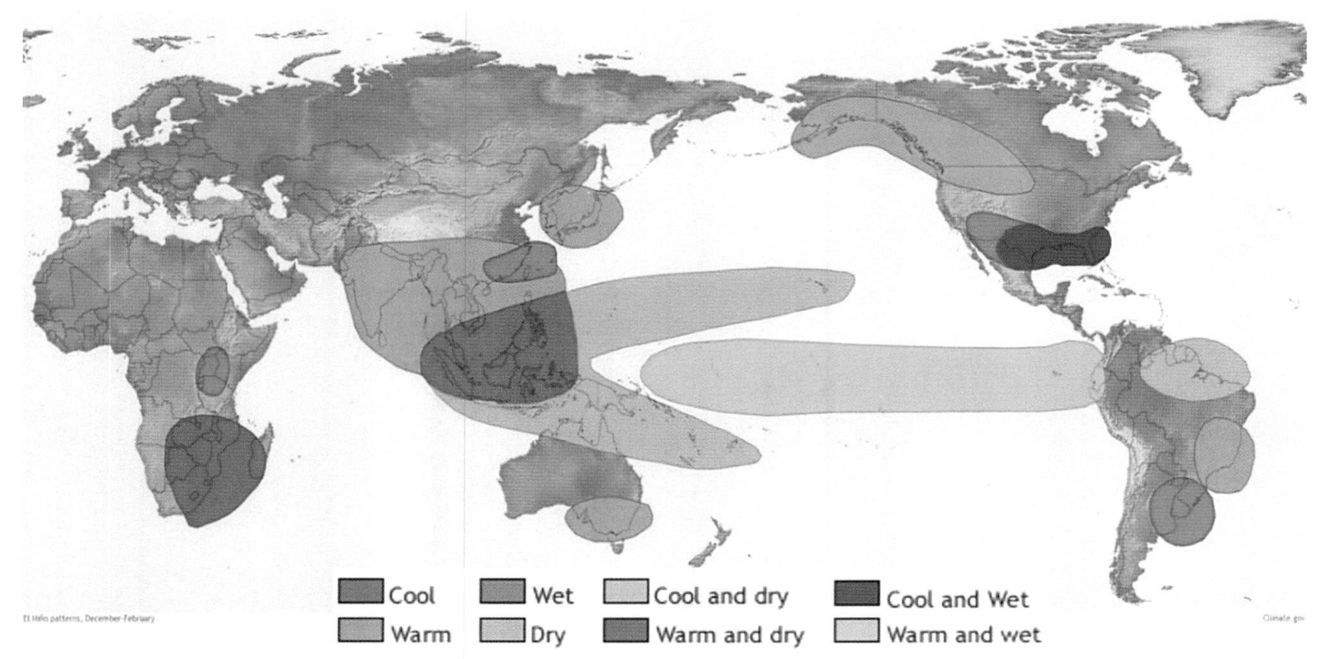

Fig. 9.7 Global impact of ENSO in the northern hemisphere winter months. *(Modified from NOAA Climate.gov.)*

El Niño drought and the reverse on the Americas coasts. The cycle is known as the El Niño Southern Oscillation (ENSO) and causes severe disruption in way of life, including fisheries and agriculture.

Whereas the cause of ENSO is temperature change in the equatorial Pacific, the effects are more widespread. Fig. 9.7 shows the results in the northern hemisphere winter months of December to April. In Australia, where these are summer months, the effect is to make some regions in the southeast hotter than normal. The fire-prone summer months are now rendered even more dangerous. Vast portions of southeast Asia and east Africa experience drought and extreme heat. The map underlines the fact that El Niño simultaneously causes droughts and drenching rain in different parts of the world.

The El Niño of 1996–97 that we referred to earlier in the body of this chapter fell under the designation of a "super" El Niño, with exceptional severity. A similar one occurred in 2015–16 and was preceded by a period of extreme heat. 2011–15 was the hottest 5-year period on record. Most climate scientists believe that this was caused by the greenhouse gases introduced into the atmosphere by humans. This connection, plus the belief that the warming is certain to get worse, is the basis for the prediction that ENSO events will be more frequent and more severe. But this view lacks consensus. Some models even predict a reduction in ENSO severity and frequency. Part of the reason for the uncertainty is that ENSO data were collected for only a couple of decades, so the body of evidence is sparse in terms of long-term behavior. Whereas there is no disagreement that ocean temperature will continue to rise, the precise effect on ENSO needs more data for models to predict the complex interplay between oceanic and atmospheric processes. The jury is still out on how anthropogenic climate change and El Niño affect each other.

The western United States is a good proxy for the effect of climate change on wildfires. Much of it is arid, with precipitation primarily in the winter months. California relies on snowmelt at higher elevations as the source of water in summer. Future projections of climate in the region indicated warming and reduced snow at moderate-to-high elevations. Early snowmelts could be expected. Westerling examined data from 1980 to 2003 and concluded that the largest fires were in early snowmelt years.[2] A later study generally confirmed this.[16] The data shown in Fig. 9.8 are for the four decades from 1973. The correlation of areas burned with date of snowmelt is strong, generally confirming the earlier hypothesis.[2] The fire seasons in the decade 2003–12 averaged 84 days more than those in 1973–82. The average number of days of large fire (>400 ha) burn time grew from 6 in 1973–82 to over 50 in 2003–12.

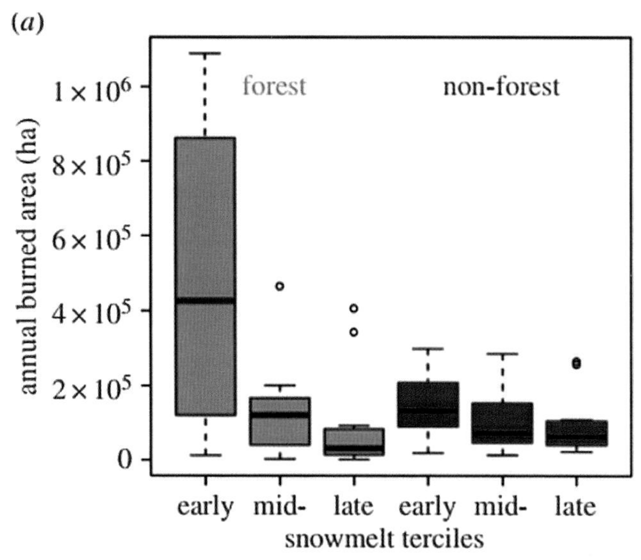

Fig. 9.8 Forest fire severity as a function of early snowmelt (1973–2012).[16] *(From Westerling AL. Increasing western US forest wildfire activity: sensitivity to changes in the timing of spring.* Philos Trans R Soc B. *2016;371:20150178. https://doi.org/10.1098/rstb.2015.0178 (web archive link).)*

In summation, wildfires are a relatively small contributor to PM in the air but are likely going to increase with further global warming. Even if strong efforts are made to reduce anthropogenic PM emissions, this source will continue to grow and become an increasingly high contributor to the total global PM inventory.

Close-up: Australia is on fire

Out of context, this title could be about cricket. Not the Jiminy sort, but a sport with fielding positions named *extra cover* and *silly point*. The Aussies are good at that sport and play it without giggling about silly point. Sadly, this is not about cricket. This is about much of the eastern portion of the continent, roughly twice the size of Belgium, being on fire at this writing in early 2020. The worst ever, by many accounts. Climate change is blamed by many but not by Prime Minister Morrison. He is also (dis)credited with vacationing in Hawaii during the crisis. I (VR) did not check original sources for this assertion, hoping that the *New York Times* story on the subject[17] did so. In 64 AD, Emperor Nero was famously said to have fiddled while Rome burned. Awkwardly, though, the class of instruments to which belongs the fiddle was not invented until 10 centuries later.

Cobargo is a small town (the *New York Times* called it a village) in New South Wales; population 900. Located on the Princes Highway, a well-traveled route going north-south (think State Route 1 in California, with two lanes much of the way), it is a stone's throw from Bermagui, a fishing village transformed into a tourist haven. Home to the best marlin fishing in the world, according to no less an authority than Ernest Hemingway. It is also where my sister-in-law Margaret calls home. Fires were in scores of locations, and Cobargo clocked in at a hot 43°C (about 110°F) with high winds. Concern ran high, and then came the order to evacuate. That was rescinded almost immediately because the three roads out of town were engulfed in spots and impassable. Cars were required to be stocked and ready to roll, awaiting the order to pull out. Power and water went down, and communications ceased to the outside world. Being small became a virtue for local communication. Cell phones could only be charged on a car battery, and cars were advised not to be started up except to evacuate.

Then the fire front swept through. For some reason, this town caught the attention of the news helicopters, and we, waiting helplessly in the United States, saw a before-and-after image of the main street, a stretch of the Princes Highway. The west side of the street with the coffee shop and bakery, the shop from which I bought handsewn leather, the little grocery store—all gone. The east side, completely intact. Seemingly; after all, it was just one still image. Straws get clutched at in these instances. Ours was that Margaret's house was three blocks *east* of the main street. In the end, our optimism was justified.

Margaret was raised on an inland farm with annual rainfall of 8 in. And no electricity in the village until well into her teenage years. In areas such as these, almost all the water used in the home came off the roof. Folks from those parts make do with what they have. The impossible simply takes longer. So it was that in Cobargo, with 40 in. of rainfall in normal years and city water piped in, she still captured rainwater. When the power and water went down, she had water, for most who needed it. These communities are resilient, and everybody does their part. At least that would be what she would say, if asked.

The Prime Minister, he of the Hawaii sojourn, came to visit. As noted above, for some reason this little hamlet had drawn attention. When he attempted to shake the hand of a woman resident, she informed him that she would shake it only if he increased funding for firefighters (he was credited with disallowing an authorization of overtime).

"You won't be getting any votes down here, buddy," said a man. Yet another said, "You're out, son."[17] Ordinarily, rural areas such as this are reliably conservative (Mr. Morrison is from a conservative party). Not this time.

At this writing, power, water, and communications are all back on in Cobargo. People are catching their breaths. Even smiling grimly about it.

The bookstore has a sign outside: "Post-Apocalyptic Fiction has been moved to Current Affairs." But the sobering realization is that the fire season still has at least 6 weeks to go.

References

1. World Health Organization. *Household air pollution and health fact sheet*; May 8, 2018. http://www.who.int/en/news-room/fact-sheets/detail/household-air-pollution-and-health.
2. Westerling AL, Hidalgo HG, Cayan DR, Swetnam TW. Warming and earlier spring increase western US forest wildfire activity. *Science*. 2006;313:940–943. https://doi.org/10.1126/science.1128834.
3. Okoshi R, Rasheed A, Reddy GC, McCrowey CJ, Curtis DB. Size and mass distributions of ground-level sub-micrometer biomass burning aerosol from small wildfires. *Atmos Environ*. 2014;1352-2310. 89:392–402. https://doi.org/10.1016/j.atmosenv.2014.01.024.
4. Kleeman MJ, Schauer JJ, Cass GR. Size and composition distribution of fine particulate matter emitted from wood burning, meat charbroiling, and cigarettes. *Environ Sci Technol*. 1999;33(20):3516–3523. https://doi.org/10.1021/es981277q.
5. Johnston F, Hanigan I, Henderson S, Morgan G, Bowman D. Extreme air pollution events from bushfires and dust storms and their association with mortality in Sydney, Australia 1994–2007. *Environ Res*. 2011;111(6):811–816. https://doi.org/10.1016/j.envres.2011.05.007.
6. Ford B, Val Martin M, Zelasky SE, et al. Future fire impacts on smoke concentrations, visibility, and health in the contiguous United States. *GeoHealth*. 2018;2:229–247. https://doi.org/10.1029/2018GH000144.
7. Rutledge D. Estimating long-term world coal production with logit and probit transforms. *Int J Coal Geol*. 2011;85(1):23–33. p. 32 https://doi.org/10.1016/j.coal.2010.10.012.
8. Wegesser TC, Pinkerton KE, Last JA. California wildfires of 2008: coarse and fine particulate matter toxicity. *Environ Health Perspect*. 2009;117(6):893–897. https://doi.org/10.1289/ehp.0800166.
9. Kim YH, Warren SH, Krantz QT, et al. Mutagenicity and lung toxicity of smoldering vs. flaming emissions from various biomass fuels: implications for health effects from wildland fires. *Environ Health Perspect*. 2018;126(1):2018.
10. Urbanski S. Wildland fire emissions, carbon, and climate: emission factors. *For Ecol Manage*. 2014;317:51–60. https://doi.org/10.1016/j.foreco.2013.05.045.
11. Doerr SH, Santín C. Global trends in wildfire and its impacts: perceptions versus realities in a changing world. *Philos Trans R Soc B*. 2016;371(1696). https://doi.org/10.1098/rstb.2015.0345.
12. Syphard AD, Keeley JE. Location, timing and extent of wildfire vary by cause of ignition. *Int J Wildland Fire*. 2015;24(1):37. https://doi.org/10.1071/WF14024.
13. Ganteaume A, Guerra F. Explaining the spatio-seasonal variation of fires by their causes: the case of southeastern France. *Appl Geogr*. 2018;90:69–81. https://doi.org/10.1016/j.apgeog.2017.11.012.
14. Penn I, Eavis P. PG&E pleads guilty to 84 counts of manslaughter in camp fire case. *New York Times*. June 18, 2020. https://www.nytimes.com/2020/06/16/business/energy-environment/pge-camp-fire-california-wildfires.html. Retrieved 20 June 2020.

15. San Francisco Chronicle; 2017. https://www.sfchronicle.com/bayarea/article/Underground-power-lines-don-t-cause-wildfires-12295031.php. Retrieved 29 January 2020.
16. Westerling AL. Increasing western US forest wildfire activity: sensitivity to changes in the timing of spring. *Philos Trans R Soc B*. 2016;371. https://doi.org/10.1098/rstb.2015.0178.
17. New York Times; 2020. https://www.nytimes.com/2020/01/03/opinion/australia-fires-climate-change.html?smid=nytcore-ios-share. Retrieved 23 January 2020

Reducing PM and NO_x in diesel engine exhaust

Chapter 2 discussed the general principle of operation of internal combustion engines. Statistically, diesel engines, comprising transport vehicles such as cars, buses, locomotives, trucks, and vessels, are the largest contributors to particulate matter (PM) in urban air in the developed world.[1] They are significant contributors in cities in developing economies. Additional sources, although smaller, are backup electric power systems, farm machinery, and electricity generators, where there is no grid. This chapter will discuss the means by which PM and contributory pollutants such as NO_x and sulfur compounds are ameliorated. The means fall into three general classes: modifications to the engine, treatment of the exhaust gases, and modifications to the fuel consumed.

Typical diesel engines run on a four-stroke cycle. Air is introduced into the combustion chamber in cycle 1. In cycle 2, this air is compressed to the desired extent. Near the end of this cycle, fuel is injected into the chamber. The act of compression in cycle 2 produces frictional heat, and, if sufficiently compressed, the injected fuel will ignite. This is known as compression ignition, as opposed to spark ignition in a gasoline engine. When diesel is substituted with natural gas or liquefied natural gas, a spark is required. The explosion drives the piston that, in turn, provides motive power to the wheels. This is cycle 3. The spent gases are exhausted in cycle 4, and the system returns to cycle 1. Interventions to improve emissions performance address one or more of these cycles.

In gasoline engines, the fuel and air are usually mixed prior to injection into the combustion chamber—in commonplace terminology, the cylinder. Before about 1992, this mixing was usually done in a carburetor. Fuel injection technology changed that, and it is done in one of two ways. Ported injection is very similar to carburetion, in that the fuel and air are mixed before injection. In direct injection means, the fuel goes directly into the cylinders. Today's gasoline engines use multiple injections within a cycle, based on cylinder temperature. Although this has many advantages with respect to evaporative cooling, one drawback is that the fuel mixture with

Particulates Matter
https://doi.org/10.1016/B978-0-12-816904-9.00005-2

air is inhomogeneous. This can result in unburnt hydrocarbons. However, design of the injection systems can ameliorate the effects. The older, lower cost means uses a cylinder wall-guided system, wherein the fuel–air mixture impacts the piston head and forms a cloud that then transports to the spark plug and ignites. This leads to inhomogeneities. The newer, costlier means is injection in a centrally mounted nozzle, forming a hollow cone. The resulting cloud remains reasonably well mixed and stable until ignition.

Inhomogeneous mixing is present in virtually all diesel fuel injection designs. The evidence for this is that the downstream treatment remedy for unburnt hydrocarbons, the diesel oxidation catalyst (DOC), is now standard on most engines. Local volumes have lower oxygen-to-fuel ratio (below stoichiometric) resulting in fuel not combusted. Therefore the likelihood of unburnt hydrocarbons (organic molecules) is greater in compression ignition engines. Typical classes of organics include aliphatic and aromatic hydrocarbon molecules, both saturated and unsaturated. Typical molecules include aldehydes, such as formaldehyde, benzene, butadiene, and polycyclic aromatic hydrocarbons. These organics react with the black carbon or other nuclei (such as metal droplets) in the PM produced and make it more toxic.[2] These reacted particles are generally referred to as primary aerosols, in that the reactions occur with coproduced organics. Later, in the atmosphere, organics present from other sources are responsible for the production of secondary aerosols, often mediated by photochemical action by the photons derived from sunlight.[3] As a point of note, most evaluation of diesel emissions is conducted at the tailpipe. Accordingly, secondary aerosols are not adequately represented in such studies. However, even the tailpipe studies are likely relevant to populations such as railway workers and automotive repair shop personnel. Street canyons, which can loosely be defined as streets bounded by tall buildings, may also have a high proportion of primary aerosols in the immediate vicinity.[4,5]

Diesel particulate filters

The principal design means for minimizing particulate emissions from diesel engines is the diesel particulate filter (DPF). All versions have in common the ceramic filter intended to capture PM. However, after the filter is loaded to a certain level, the back pressure on the system reduces engine performance by increasing fuel consumption. To avoid this, the filters are periodically cleaned. This is accomplished by a process known in the industry as regeneration, so named because the filter efficacy is revived. The process simply is to remove the deposit by combusting it. The means vary, but they all attempt to convert the carbon to CO_2. There are some unintended consequences that will be discussed later in this chapter. These devices, downstream of the combustion chamber, can in principle be disposable or regenerative. Most commercial vehicular systems use regeneration and that is the only design discussed here.

Particulate filters are all constructed from ceramic materials such as silicon carbide, aluminum titanate, and synthetic cordierite, the last being a mixed oxide of Mg, Al, and Si. All these materials have good thermal shock properties, although in the case of cordierite, the manufacturing process must take care to align the crystals to minimize thermal expansion along one axis. The need for ceramics is in part the high temperature of the exhaust and in part to tolerate the imposed higher temperature during the regeneration step. They have porosities in the range of 40%–60%, and the porous network comprises pores of 10–30 μm in diameter.

The channels terminate at the ends. This forces the gases to enter the porous ceramic walls. The high porosity and permeability allow easy transport. Within the ceramic walls, the particles permeate through a diffusion mechanism. This carries on until carbon begins to form a layer on the outer surface of the ceramic. When this occurs, the back pressure increases, causing engine efficiency loss. The back pressure increases in a nonlinear fashion because the mechanism of clogging with soot changes over time. This adds a measure of complexity in using back pressure as a guide to the triggering of the active systems.

Regeneration may be required as often as every 500 km. The process can be active or passive. Passive systems rely on oxidation of the carbon by reactants present in the off-gases, sometimes aided by a catalyst. One means is to use nitrogen oxides. NO_2 will oxidize carbon at temperatures exceeding 250°C. This is aided by the addition of an oxidation catalyst into the system. Catalytic action can also first oxidize NO to NO_2 and then allow the NO_2 to do its job. The catalyst can also cause direct oxidation of the soot by O_2. The catalyst may be introduced into the fuel as a catalyst precursor. It can also be directly placed on the filter. A third means is to place an NO_2-producing catalyst just upstream of the filter.

Active means achieve the soot burnoff by increasing the temperature of the soot to allow for some of the reactions mentioned above. The simplest approach is to modify engine-operating conditions to produce hotter gases. One such would be to make late cycle introductions of fuel. A second means is to produce an exothermic reaction upstream of the filter. Fuel may be combusted in a burner or by catalytic combustion. A further means is an electric heater either upstream of the filter or incorporated into the filter. General Motors introduced one such in 2013, and the associated patent claims to reduce the regeneration time from 20–30 min to 4–6 min.[6] The reduction in regeneration time is of interest because the particle count of emissions during this period can be orders of magnitude greater than in normal operation, as discussed later in this chapter. The morphology of the emitted particles may also be different for the rapid combustion process.

Yet another means is the placement of a DOC upstream of the DPF. One function of this device is to oxidize NO to NO_2 that is an oxidant in the DPF. The exothermic reactions in the DOC also serve to heat the gases, which is useful for DPF regeneration functions. All the foregoing are energy intensive, either in the consumption of more fuel or use of electricity.

Finally, a combination may be used. The passive addition of a DOC can reduce the temperature increase needed in the active means. This could reduce the energy consumption in the active system or reduce the duration of the regenerative step. These benefits are traded against the cost of the introduced catalyst.

In general, regeneration proceeds until the back pressure is down to prescribed limits. However, because back pressure changes are nonlinear, the trigger is likely reliant on some measure of data analytics.

Regeneration results in higher nanoparticle count

Because filter regeneration involves combustion of carbon particles (with or without a layer of organic molecules), one ought not to be surprised by the creation of new particles. In some ways, it is not so different from combustion of finely particulate coal. Ash is produced as well, although not to the extent with coal, because of fewer inert oxides. The ash does have to be removed periodically.

Undoubtedly, the total mass of particles originally emitted from the engine is reduced. But the combustion produces ultrafine particles that may then grow. The reported size of the nucleating mode particles is in the 10–20 nm range. The total number of particles can be orders of magnitude greater than the number during normal engine operation.[7]

Another point of interest is that the filter cake initially formed is believed to act as a filter for the ultrafine particles. Removal of this cake by combustion will reduce the total mass of particles but is reported to result in larger numbers of particles than in the preregeneration phase. In one study, regeneration reduced mass of emissions by a factor of two, but the particle count went up 20 times.[8]

Diesel oxidation catalyst

The primary function of this device is to combust the unburnt hydrocarbons and CO. But it is also used to oxidize the NO to NO_2, to be an oxidant in the DPF. The reactions are

$$CO + ½\,O_2 \rightarrow CO_2 \tag{10.1}$$

$$NO + ½\,O_2 \rightarrow NO_2 \tag{10.2}$$

$$C_3H_6 + 9/2\,O_2 \rightarrow 3\,CO_2 + 3\,H_2O \tag{10.3}$$

The catalysts used are noble metals such as platinum (Pt), palladium (Pd), or rhodium (Rh). The primary means for assuring high surface area for the catalyst elements is the use of oxides of aluminum, cerium, and zirconium in combination. This also serves to minimize grain-to-grain contact of the catalyst, thus reducing the likelihood of sintering of the catalyst elements, which would cause a diminution of catalytic activity. Pt and Pd are the most common catalysts used. The carrier structure is a honeycomb monolith of ceramic or metal. Virtually, all modern diesel engines use DOCs and achieve reductions in hydrocarbons and CO of up to 90%. The exotherms in the DOC serve to automatically provide the higher temperatures needed for regeneration in the DPF, and with a DOC present, the regeneration alternatives are narrowed. Interestingly, the DOC will also cause the organic layer on carbon nuclei to be oxidized, rendering it less toxic. The organic layer is formed because the nanoparticulate nucleus, typically less than 20 nm in size, has a very reactive surface and readily adsorbs coproduced organics. Because the organic layer on carbon is believed to be disproportionately (when compared with the carbon core) harmful to health, the DOC performs a useful function relative to the fate of the particles. However, secondary organic deposition mechanisms, from organics in the atmosphere, will still be in play.

NO_x removal

Nitrogen in the combustion air in the cylinder will oxidize to one of the oxides, NO or NO_2. Engine conditions determine the degree to which this happens. The two principal variables are temperature and oxygen concentration. These are not independent variables because a lean mix—hydrocarbons below stoichiometric levels—will burn hotter. In this condition, the oxygen is relatively higher in concentration. Lean mixes are more efficient in fuel utilization and give better engine performance. The higher exhaust temperatures benefit the downstream processes in the DOC and the regenerative step in the DPF. In an ironic twist, the higher temperatures associated with this high NO_x production condition also benefit one of the NO_x reduction means, selective catalytic reduction (SCR). The other means, the lean NO_x trap (LNT), requires bursts of the opposite, rich mixture burning; more on that later. It seems that in attempting to balance fuel efficiency and emissions, some piper is always getting paid. Unless one cheats. This tussle was certainly present when Volkswagen (VW) fiddled with the LNT operation. More on that later as well.

Exhaust gas recirculation

This is the simplest, in-engine means for NO_x control. The gases exhausted from the cylinder are collected and reinjected into the cylinder. Sometimes a heat exchanger is used to cool down the gases prior to injection. The primary purpose is to reduce the overall oxygen loading in the chamber. This causes a lower temperature burn that in turn reduces the oxidation of the nitrogen in the air mix. The net result is lower output of NO_x. The unintended consequence is uncombusted hydrocarbons and CO. On balance, the results must be considered positive because the industry has chosen this system to be in common use.

The Jake Brake

Who among us has not heard a staccato burst from an 18-wheeler decelerating on the highway? In the event you have pondered the origins of this intrusive sound, you have now come to the right place. The rapid-fire sound is a consequence of application of a device colloquially known as a Jake Brake. In short, it is an engine braking means used by trucks to avoid using friction brakes, for reasons of cost.

Ordinarily, fuel is injected at the top of the compression cycle, and the heat of compression ignites it, using the air that has been compressed, which drives the piston down to power the wheels. A Jake Brake system is deployed by a manual switch in driver control. It deliberately opens the exhaust valves at the top of the compression cycle, venting the high-pressure gases. The pressure is lowered significantly, and any fuel present will not ignite at the lower pressures and temperatures. The downstroke will not provide power to the wheels. In effect, it is providing a braking action. This action produces a staccato sound that is very recognizably loud and piercing. According to the manufacturer, special mufflers on the truck can ameliorate this.

According to the lore, Clessie Cummins, founder of the Cummins Engine Company, had a near disaster with failed brakes on a California mountain downhill segment in 1931. By 1954, he had invented a device to supplement the braking using the engine. But, for some reason, it was not initially implemented in Cummins engines and instead was taken up by Jacobs Engineering. It was marketed by them as the Jacobs Brake. Put into wide use, truckers took to referring to them as Jake Brakes.

Communities can, and do, put up signs disallowing the use of Jake Brakes in residential areas and certainly in noise-abatement areas. Restrictions are common in hilly sections. Because the operation is manual, the restriction could easily be followed by truckers. However, they might feel secure in ignoring the sign, because, as in speeding, the transgression must be proven. In this case, the cop would need a specialized sound-recording device, which is unlikely to be standard issue. The Jacobs folks likely had been delighted by the familiar use of the term "Jake" in generic form. The enthusiasm was no doubt dampened when it was used in signs such as "No Jake Brakes in city limits." The publicity sword is double edged.

Selective catalyst reduction

Ammonia is used as a reducing agent to convert NO_x to gaseous nitrogen and water. The reactions are

$$4\,NO + 4\,NH_3 + O_2 \rightarrow 4\,N_2 + 6\,H_2O \tag{10.4}$$

$$6\,NO_2 + 8\,NH_3 \rightarrow 7\,N_2 + 12\,H_2O \tag{10.5}$$

$$2\,NO + 2\,NO_2 + 4\,NH_3 \rightarrow 4\,N_2 + 6\,H_2O \tag{10.6}$$

Eq. (10.6) is more energetically favored and has a higher rate of reaction than the other two. This would appear to make it preferable to not have a precursor DOC that would convert much of the NO to NO_2 and require the reaction in Eq. (10.5), with a resultant diminution in the efficiency of conversion of the NO_x. However, proper design of the DOC can minimize this deleterious effect, and the practice is to shoot for a NO/NO_2 ratio of about 1.1.

The ammonia is supplied in the form of urea $(NH_2)_2CO$ combined with water to about a third mass fraction of urea. The solution is sprayed onto the exhaust gases, the water evaporates, and the urea decomposes to ammonia and isocyanic acid by the reaction:

$$(NH_2)_2CO \rightarrow NH_3 + HNCO \quad (10.7)$$

The isocyanic acid further hydrolyzes to more ammonia by the reaction:

$$HNCO + H_2O \rightarrow NH_3 + CO_2 \quad (10.8)$$

Each molecule of urea produces two molecules of ammonia for the SCR reaction. Whereas other means for delivering ammonia exist, this is the commercially popular method, known as AdBlue in some countries.

Exhaust gas temperature is a key to proper operation. The temperatures must be in excess of 200°C prior to injection of the urea mixture. On the upper end, the temperature is limited to 600°C, above which the ammonia burns prior to the intended SCR reactions occurring.

The key drawback of this method is the need for a urea container, injection system including a pump and control box, the space they occupy, and the need to fill the container periodically. In trucks, none of these is an issue, in part because the refill procedure is built into the truck operation. In passenger vehicles, space is a constraint, especially in smaller cars. Customer objection to the nuisance of refilling was another factor in holding back the introduction of this technology. The LNT method is believed to have been invented to address this customer predilection, especially for the US market, where the penetration of diesel passenger vehicles had been low until about 2008, when VW made a big push, using the LNT. Therein, lies a tale unto itself.

Lean NO_x trap

There are two steps (there is a preliminary step that we will skip for simplicity). In the first, NO_x is captured on a coating that *adsorbs* NO_x. Adsorption is a surface phenomenon that is easily reversed. The standard means for reversal is either pressure or temperature. Pressure swing absorption, as the process is known, involves adsorption at a certain pressure, usually high, followed by a change in pressure that desorbs the catalyst. Temperature swing adsorption similarly uses temperature changes to adsorb and desorb. In these cases, the desorption produces the same species adsorbed but now in concentrated form. In the first step, it may have been present in dilute form. A commercial example of this is CO_2 capture from integrated gasification combined cycle coal fired plants, as discussed in

Chapter 13. In that case, a relatively pure stream of CO_2 is collected for sequestration or other use.

In the LNT process, the desorption is activated when the coating is considered filled up. This involves removing the NO_x to regenerate the coating activity. Unlike in swing adsorption, the desorption is achieved through a chemical reaction that is energetically favored over leaving the NO_x adsorbed. The NO_x is reduced to nitrogen and CO_2 on a special catalyst by reacting it with some mixture of hydrocarbons, hydrogen, and CO. This mixture is created by switching the engine to a rich burn mode, away from the lean. The reactant is the fuel from the cylinders that is only partly combusted. Not surprisingly, during that time, engine performance drops because the fuel is not fully utilized for producing power. Fuel mileage is reduced, as is torque. In principle, the hydrocarbon could be introduced separately into the exhaust upstream of the trap. Hydrogen could also be stored and injected, but this is considered potentially hazardous. Accordingly, running the engine rich is the only common means used. This is the key step that got VW in trouble.

How the lean NO_x trap got sprung

The beginnings were innocuous. The International Council for Clean Transportation (ICCT) funded a small study at West Virginia University to quantify the NO_x emissions from European diesel vehicles introduced into the United States a few years prior. On-road NO_x emissions had anecdotally been reported as worse than in stationary tests. The regulations in the United States were more stringent than in Europe. ICCT hoped that the study would quantify the advances made by German manufacturers to address the US market. In a huge marketing push ("It's not your grandfather's diesel") in 2008, VW promised fun and economical driving combined with excellent emission controls. Nothing understated about this launch; advertisements were run at the Super Bowl. The ICCT hoped to persuade manufacturers to offer the advances in European vehicles. Possibly presciently, the study required on-road testing, in addition to standard emissions testing in the shop.

Enter graduate students Marc Besch and Arvind Thiruvengadam at the Center for Alternative Fuels Engines and Emissions in the West Virginia hills. This was a lightly funded portion of West Virginia University and glad of the contract. The Center had developed a system to perform emissions testing on the road in passenger vehicles, which is likely the reason they won the contract with the ICCT. Regulatory compliance emissions testing is almost always conducted in the shop; this represented a departure and was probably intended simply to elucidate the differences under different driving conditions. Arvind and Marc, foreign students respectively from India and Switzerland, were gearheads who loved to drive and to understand what they drove. In a display of self-admitted overkill, they drove three vehicles for thousands of miles each: a 2011 Volkswagen Jetta, a 2012 Volkswagen Passat, and a BMW X5. Hardcore enthusiasts have massive growling exhaust systems. These guys had a superstructure above the bumper that was a noiseless attention grabber. One bit of less than welcome attention drawn, although happily traversed, was that from a cop. I do not know if they pulled the foreigner card. I did once, in Australia. When pulled over, I addressed the cop as "officer," serving notice of American ignorance of whatever transgression was to be alleged. The conversation went something like this:

> *Policeman: Sir, did you recently purchase fuel? [This is how they talk Down Under.]*
> *A Much Relieved Me: Oh, did I leave the gas cap off? [Note the clever use of "gas," not "petrol."]*
> *(unsmiling) Policeman: Sir, did you pay for the fuel?*

So of course, I turned to my wife, who, being Australian, ought to know quaint rules such as paying for stuff. "Did you pay for the gas when you went into the convenience store to buy the other things?" Needless to mention, there was no credit card payment at the pump in this far outpost. Also, the response was not comforting, at multiple levels, mostly centered on the undeniable fact that it was I who had filled the gas tank. He required us to return to the previous town to pay. Said we would be met by his colleague at the town limit. Only one road back, and no side roads, so no worries about decamping without paying. An offer to call in a credit card number was not even dignified with a response. Back we went, thankfully only 35 km, and no cop was in evidence at the town limit. He was in the shop, flirting with the paid help. Paid no attention to us whatsoever.

Arvind and Marc's results were stunning and inexplicable for the Jetta (Vehicle A) and Passat (Vehicle B). The BMW X5 (Vehicle C) performed as expected, within regulatory limits. It carried a urea-SCR that performed satisfactorily, except for one rural run. The other two exceeded the regulatory limit, a standard that they had passed in the shop (designated by the "bag 3" dynamometer testing). The Jetta emissions were as much as 35 times greater than permissible. In both vehicles, highway results were better, but there was no difference between rural hills and dales when compared with city driving. The students searched for a hardware explanation and found none. Everything was double and triple checked. There were excursions into self-doubt. But the results were solid. No reasonable explanation made sense. They decided to simply inform the California Air

Resources Board (CARB) and let them sort it out. CARB conducted tests, as did the US EPA, and lawsuits were filed. An unlikely David in the West Virginia hills had inadvertently stepped on the toes of a Goliath.

Eventually, VW admitted to using a "defeat device." While the details were not mentioned in the guilty plea, the phenomenological feature was clear. The software was able to recognize when the engine was on conventional emissions testing. In principle, this could be from any number of triggers, such as the hood being raised. The precise triggers were not revealed, either by the company or by the individual engineers who pled guilty to the falsification. Perhaps, nobody really cared to know. When the control system believed that the engine was in test mode, the full LNT features kicked in, with the engine running rich, and NO_x emissions were satisfactorily low due to reduction in the NO_x to N_2 by the reducing gases. In normal road operation, it defeated the system by switching off this key LNT feature, and the engine ran lean continuously, providing superior torque and low fuel consumption. But the NO_x emissions jumped up, exactly as expected in an oxygen-rich combustion mode, as shown in Fig. 10.1.

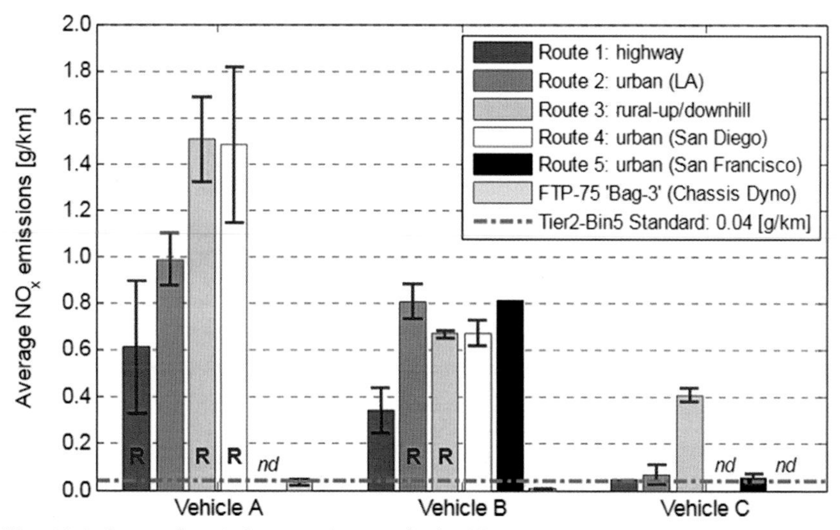

Fig. 10.1 On-road emissions testing results by West Virginia University. Average NO_x emissions of test vehicles over the five test routes when compared with the US Environmental Protection Agency (US-EPA) Tier2-Bin5 emissions standard; repeat test variation intervals are presented as $\pm 1\sigma$; Route 1 for Vehicle A includes rush-hour/nonrush-hour driving; "R" designates routes including a test with DPF regeneration event; "nd" indicates no data are available. *(Courtesy Arvind Thiruvengadam and Marc Besch, West Virginia University.)*

One interesting aspect to the episode is that the term "defeat device" already existed in the industry lexicon. But the sheer scale and audacity were astounding. This was not just some software that got installed, transparent to all but a few. People had to verify the defeat aspect while operational and under test, for every model. Technicians had to know what was going on. The targets were the smaller vehicles. Small car engines (2 L, for example) can ill afford the loss of torque associated with running rich. More particularly, it erodes the fuel economy advantage of diesels. These cars are bought largely for the fuel economy, so impairment in that area is a potential showstopper. The smaller engines also have less room for an SCR device, even if one were to ignore the customer dissatisfaction with having to refill the urea container. These appear to be the motivations for the deceit.

A sad story of the little engine that could but never got the chance.

Gasoline engines: Are they getting a free pass?

The principal features of gasoline engines were discussed in Chapter 2. With the widespread use of the three-way catalytic converter (TWC) in the 1980s, emissions in general were reduced substantially. Gains were also made when lead was removed. Lead had been added to gasoline as an octane booster, in the form of tetraethyl lead (TEL).

The TWC removes the three principal pollutants, NO_x, CO, and unburned hydrocarbons, in a single chamber attached to the exhaust system. It comprises a metal or ceramic honeycomb structure with the catalyst coated on the surface. The purpose of the catalyst is to reduce the temperature needed for the desired reaction. The reactions are

$$2NO \rightarrow N_2 + O_2 \tag{10.9}$$

$$2NO_2 \rightarrow N_2 + 2O_2 \tag{10.10}$$

$$2CO + O_2 \rightarrow 2CO_2 \tag{10.11}$$

$$C_3H_6 + 9/2\,O_2 \rightarrow 3\,CO_2 + 3\,H_2O \tag{10.12}$$

The first two reactions typically occur in the first stage of the converter, where reducing catalysts are present on the honeycomb surfaces. The second stage has oxidation catalysts to convert CO and hydrocarbons (one example shown in Eq. 10.12) into CO_2 and water. The heat of combustion in the engine cylinders is the source of the temperature needed. Cold starts necessarily result in poor initial performance of the device.

Octane is a pure chemical with $n = 8$ in the formula C_nH_{2n+2}. Gasoline is said to have an octane rating of 87 (also known as regular), when it has a mixture of compounds that approximate the performance of a mixture of 87% octane and 13% heptane. The performance in question is the property of not autoigniting under-designed compression ratios (see Chapter 2). With additives, the octane rating can be increased to 91 and more. The first commercial additive was TEL, introduced in the 1920s. This caused lead to be present in the exhaust emissions. While this was known, and lead had already been removed from paint in the middle part of the twentieth century, the use of TEL continued until the introduction of the TWC. Lead fouled the catalytic converter, which had proven invaluable and had been broadly accepted by the automotive industry. The primary impetus for the law changing in 1995 to ban lead in gasoline for consumer vehicles was the impact on the TWC, not atmospheric lead in emissions.

Particulate emissions were not generally believed to be an issue with gasoline engines. Diesel engines billowed black smoke and were the visible bad actors. However, recent research has demonstrated that gasoline engines do produce ultrafine particles, on par with diesel engines with the full battery of control measures discussed earlier in this chapter. This is evident when the metric employed is particle count, rather than mass. Fig. 10.2 shows results from a study by Mohr et al.[7]

The emissions measured by particle count are generally similar for gasoline and diesel engines, when the latter have the full complement of control equipment, especially the DPF. As expected, diesel emissions are worse during and shortly after regeneration of the DPF. Unsurprisingly, the greatest emissions are in the diesel engine without a DPF (the gray bar in Fig. 10.2). Interestingly, direct injection gasoline engine performance is worse on PM count when running lean (the two VW Touran green bars). Direct injection is favored because of the higher combustion efficiency. As discussed in Chapter 2, it can also take advantage of evaporative cooling to tolerate higher compression ratios. But the higher particle count appears to be a pervasive downside. Offsetting that is the fact that the count is distributed over more kilometers, if indeed the fuel economy is increased by the higher compression.

Finally, investigators have reported that gasoline engines suffer from significant increases in particle count at lower ambient temperatures. One study, conducted at −7°C and 22°C, demonstrated dramatically higher black carbon and secondary organic aerosol emissions from gasoline engines at the lower temperature, with emissions up to 400 times higher than those for the diesels, which appeared relatively unaffected even at the lower temperature.[9]

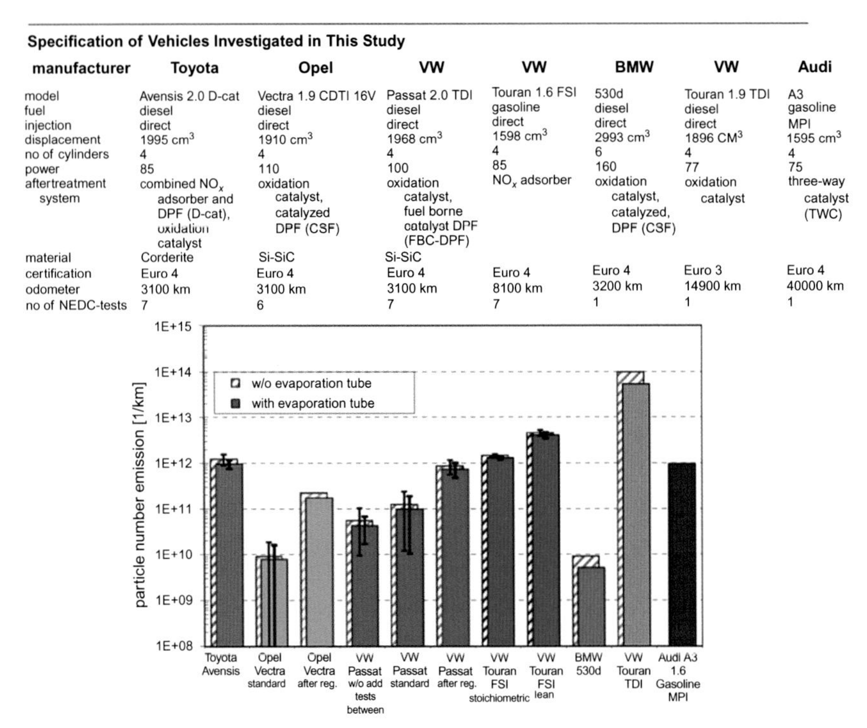

Specification of Vehicles Investigated in This Study

manufacturer	Toyota	Opel	VW	VW	BMW	VW	Audi
model	Avensis 2.0 D-cat	Vectra 1.9 CDTI 16V	Passat 2.0 TDI	Touran 1.6 FSI	530d	Touran 1.9 TDI	A3
fuel	diesel	diesel	diesel	gasoline	diesel	diesel	gasoline
injection	direct	direct	direct	direct	direct	direct	MPI
displacement	1995 cm^3	1910 cm^3	1968 cm^3	1598 cm^3	2993 cm^3	1896 CM^3	1595 cm^3
no of cylinders	4	4	4	4	6	4	4
power	85	110	100	85	160	77	75
aftertreatment system	combined NO_x adsorber and DPF (D-cat), oxidation catalyst	oxidation catalyst, catalyzed DPF (CSF)	oxidation catalyst, fuel borne catalyst DPF (FBC-DPF)	NO_x adsorber	oxidation catalyst, catalyzed, DPF (CSF)	oxidation catalyst	three-way catalyst (TWC)
material	Corderite	Si-SiC	Si-SiC				
certification	Euro 4	Euro 4	Euro 4	Euro 4	Euro 4	Euro 3	Euro 4
odometer	3100 km	3100 km	3100 km	8100 km	3200 km	14900 km	40000 km
no of NEDC-tests	7	6	7	7	1	1	1

Fig. 10.2 Particle count for diesel and gasoline engines under various operating conditions and different emission-control systems. Error bars represent ± 1 standard deviation.[7] *(From Mohr M, Forss A-M, Lehmann U. Particle emissions from diesel passenger cars equipped with a particle trap in comparison to other technologies.* Environ Sci Technol *2006;40(7):2375–2383. https://doi.org/10.1021/es051440z (web archive link).)*

The major takeaway from the recent studies, especially the ones using particle count rather than mass as the metric, is that gasoline engines may have been getting a free pass on particulate emissions. Of course, the comparisons have mostly been with diesel engines with all the bells and whistles, and legacy vehicles will continue to cause diesels to collectively be worse than gasoline vehicles.

References

1. Karagulian F, Belis CA, Dora CFC, et al. Contributions to cities' ambient particulate matter (PM): a systematic review of local source contributions at global level. *Atmos Environ.* 2015;120:475–483. https://doi.org/10.1016/j.atmosenv.2015.08.087.
2. Ebersviller S, Lichtveld K, Sexton KG, et al. Gaseous VOCs rapidly modify particulate matter and its biological effects—part 1: simple VOCs and model PM. *Atmos Chem Phys.* 2012;12(24):12277–12292. https://doi.org/10.5194/acp-12-12277-2012.

3. Laskin A, Laskin J, Nizkorodov SA. Chemistry of atmospheric brown carbon. *Chem Rev.* 2015;115(10):4335–4382. https://doi.org/10.1021/cr5006167.
4. Boddy JWD, Smalley RJ, Goodman PS, Tate JE, Bell MC, Tomlin AS. The spatial variability in concentrations of a traffic-related pollutant in two street canyons in York, UK-part II: the influence of traffic characteristics. *Atmos Environ.* 2005;39:3163–3176. https://doi.org/10.1016/j.atmosenv.2005.01.044.
5. Longley ID, Gallagher MW, Dorsey JR, et al. A case study of aerosol (4.6nm<Dp< 10μm) number and mass size distribution measurements in a busy street canyon in Manchester, UK. *Atmos Environ.* 2003;37(12):1563–1571. https://doi.org/10.1016/S1352-2310(03)00010-4.
6. FleetServ. *The Future of Diesel Particulate Filter Technology.* FleetServ; 2016. http://fleetserv.com/the-future-of-diesel-particulate-filter-dpf-technology/. Accessed 1 August 2018.
7. Mohr M, Forss A-M, Lehmann U. Particle emissions from diesel passenger cars equipped with a particle trap in comparison to other technologies. *Environ Sci Technol.* 2006;40(7):2375–2383. https://doi.org/10.1021/es051440z.
8. Barone TL, Storey JME, Domingo N. An analysis of field-aged diesel particulate filter performance: particle emissions before, during, and after regeneration. *J Air Waste Manag Assoc.* 2010;60(8):968–976.
9. Platt SM, El Haddad I, Pieber SM, et al. Gasoline cars produce more carbonaceous particulate matter than modern filter-equipped diesel cars. *Sci Rep.* 2017;7:1. https://doi.org/10.1038/s41598-017-03714-9.

CHAPTER ELEVEN

Alternative fuels

Most amelioration of harmful emissions is addressed by continuing to perform the acts that result in the emissions and endeavoring to mitigate using capture schemes. We discussed in Chapter 10 the means by which particulate matter (PM) may be reduced in the exhaust from diesel vehicles. This chapter addresses the displacement of the fuel primarily responsible for the emissions. This can be accomplished in two ways. The fuel can be partially displaced with a more benign substance or completely replaced by an alternative expected to be less polluting. An extreme example of the latter is vehicles running on electricity, described in Chapter 12. This displacement or replacement is primarily of hydrocarbons used in transportation fuel. Since the discovery of plentiful oil at Spindletop, and the resulting quartering of the price of crude oil, gasoline and diesel have been the workhorses of transportation in vehicles, trains, and boats. In fact, the low cost of the fuel was a major factor in the elimination of the first coming of electric vehicles—somewhat ironical in that a century later different drivers are likely to reverse that cycle.

Gasoline and diesel used to be solely distillation products from crude oil. Even in the lightest oils, a higher molecular weight residue remains following distillation. This is used as fuel oil in ships and for home heating or cracked to yield more gasoline or diesel. Alternatives to oil-based transportation fuel were sought for two main reasons. One is to reduce the environmental footprint. The other is especially applicable to nations that are net importers of oil and comprises an attempt to use domestic sources to substitute for some of the oil-based fuel. In this latter category, South Africa was a special case during the apartheid years. Oil embargoes forced them to produce oil substitutes from coal. The same applied to Germany during the Second World War, when the Allies blockaded incoming oil tankers, forcing Germany to adapt the Fischer-Tropsch's process to convert coal into a transport fuel. This is also the process that South Africa resorted to during their embargo, tailored to their use. To this day, that process with only minor modifications is the preferred means to produce diesel from natural gas or coal.

Particulates Matter
https://doi.org/10.1016/B978-0-12-816904-9.00023-4

Diesel substitutes

The most common diesel substitutes are biodiesel, methanol, dimethyl ether (DME), natural gas, liquefied natural gas (LNG), and synthetic diesel made primarily from coal or natural gas. All these are either blended with regular diesel or used as a replacement. In the cases of natural gas in either form, the engine must be converted to spark ignition, similar to the mechanism of gasoline engines. The use of natural gas in gaseous form is a somewhat special case, applying to industrial processes such as oilfield equipment and will be lightly discussed.

Biofuel

Biofuel has two target markets: diesel engines and gasoline engines. The principal biofuel to substitute for gasoline is ethanol. Biodiesel is the term for diesel made from plant products such as soy, rapeseed, and the Canadian variant canola. Now, it applies more broadly to any biological source, such as biogas that is a term covering methane derived primarily from animal or municipal waste. A class of raw material is termed "renewable" if tax credits are sought. Fig. 11.1 shows the full spectrum of biomass conversion methods and the associated products.

The two broad classes of processing are biochemical and thermochemical conversion. In biochemical conversion, described in the top part of Fig. 11.1, the first step is to convert the raw material to sugars. Biochemical is the simplest process because the raw material is already either a monosaccharide such as glucose or a disaccharide such as sucrose, which is a combination of the monosaccharides glucose and fructose and is commonly called sugar. The fuel product is either ethanol or a diesel substitute that blends with oil-based diesel. The most common crops for making ethanol are sugarcane and corn, and these have different processing steps. The sugarcane is pressed to extract the juice to produce a syrup or molasses. The sugar is simply fermented to ethanol, as has been the practice for thousands of years to make alcohol for human consumption. Sweet sorghum and sugar beets are just as easily processed and fermented. Sugar beets are a major source of table sugar in Europe. The diversion to fuel ethanol inevitably raises the food versus fuel debate, as it does for virtually every first-generation biofuel feedstock, especially corn.

Corn processing is a bit more complicated. First, the starch is physically separated from the rest of the kernel. The saccharides in corn are in

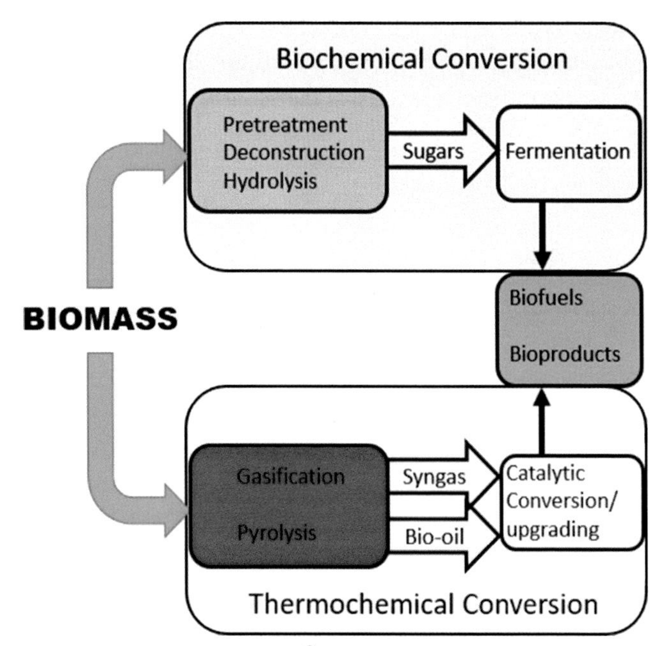

Fig. 11.1 Biomass conversion pathways.[2] *(Adapted from Dayton DC, Foust DT. Analytical Methods for Biomass Characterization and Conversion. Emerging Issues in Analytical Chemistry, vol. 10. Cambridge, MA: Elsevier and RTI Press; 2020. Courtesy: RTI International.)*

polymeric form and need to be broken down to monosaccharides and disaccharides for ease of fermentation. The process step for this is hydrolysis using amylase enzymes. These steps are simple and relatively inexpensive, which is why the United States is by far the largest producer of fuel ethanol, most of it from corn. Regulatory support is also a factor, as is the fact that presidential primaries are held first in corn states.

Thermochemical processing

Materials without ready ability to convert to sugars undergo thermochemical processing. This falls into two categories. The more generally useful for virtually any biomass is the process to produce synthesis gas, abbreviated to syngas. This is a mixture of carbon monoxide (CO) and hydrogen (H_2). The mixture will often also have some CO_2, nitrogen, and impurities that may be present in the feedstock, such as sulfur compounds. Prior to conversion of syngas to a fuel, or to any other petrochemical, it must be cleaned up with respect to the requirements of the downstream process. In the 19th century, syngas produced from coal was known as town gas and was used directly in homes for heating and cooking. When used in industrial processes, it was called producer gas. The energy content was low when compared with methane, but cheap access was a driver, with pipelines to homes and industry. This was prior to the availability of natural gas. Interestingly, before natural gas became available, catalytic conversion of syngas to methane was invented by Sabatier in 1902:

$$CO + 3H_2 \rightarrow CH_4 + H_2O \tag{11.1}$$

Shortly thereafter, in 1910, the synthesis of ammonia by the Haber-Bosch's process transformed agriculture:

$$N_2 + 3H_2 \rightarrow 2NH_3 \tag{11.2}$$

In 1923, the Fischer-Tropsch's process was invented to convert syngas to a polymerized substance approximating diesel fuel. Shortly thereafter, the synthesis of methanol from syngas was also invented. These are mentioned to underline the importance of syngas as an intermediate to several commercially important fuels. Whereas biochemical methods are relatively simple and inexpensive provided the right feedstock is available, syngas production offers the maximum flexibility for the use of the product. On the negative side of the ledger, these processes are all complex and rely on economies of scale. Recently, however, several advances have been made in conducting synthesis at smaller scales.[1] Syngas as a raw material for industrial fuels remains a workhorse, despite cheap natural gas being the raw material of choice, rather than biomass. One reason for this choice is that woody biomass has low density (Table 11.1), making the cost of transportation high. Furthermore, these processes require the ratio of H_2:CO to be at least 1.8 for methanol production and over 2.2 for Fischer-Tropsch's synthesis. Obtaining syngas from methane using steam methane reforming can produce high ratios, up to three. Biomass conversion generally yields ratios up to 1.2, and coal gasification is close to 1. Further processing known as shift reaction can improve the ratio, but clearly natural gas presents an advantage.

The high cost of transporting woody biomass spurred the final process in Fig. 11.1: pyrolysis. Note the low bulk density of wood. Coal is technically a hydrocarbon, albeit severely hydrogen challenged. Coal too is gasified to produce syngas. In fact, that is the basis for producing electricity from coal with minimum emissions. The syngas is usually shift reacted to H_2 and CO_2, with the hydrogen combusted to produce clean power and the CO_2 either sequestered or beneficially utilized.

Table 11.1 Properties of biomass when compared with pyrolytic bio-oil produced and to petroleum oil.[2]

	Wood	Coal (bituminous)	Bio-oil	Oil
Bulk density (kg/L)	0.25–0.4	0.6–1.0	1.1–1.2	0.7–1.0
Energy density (MJ/kg)	18–23	24–35	18	42
Volatile matter weight.(%)	80	15–55		
Fixed carbon weight (%)	15–20	45–85		
Ash weight (%)	1–3	3–12		
Elemental composition [weight (%), dry]				
C	50	76–90	56	84
H	6	4	6	14
O	40	10–20	38	1
N	<1	1–2	<1	1
S	<0.05	0.7–4.0	<0.05	1–3

Courtesy: RTI International.

Pyrolysis in this context is combustion with no added oxygen, resulting in breakdown of the polymers in cellulose, hemicellulose, and lignin. The reactions are depolymerization and fragmentation, and the result is a mixture of liquids that together are oil-like. As noted in Table 11.1, the bulk density is high when compared with the original woody biomass. This enables more cost-effective transport to a refining station, when compared with moving the biomass to the same station. The production of this intermediate is a key to pyrolytic processing of biomass. Some versions have the pyrolysis equipment on wheels, enabling access to relatively remote pockets of biomass. However, note also from Table 11.1 that the energy density is low when compared with conventional oil. This is consistent with the relatively low carbon content versus oil.

The most critical shortcoming of bio-oil is the high oxygen content: almost as much as in the original woody biomass. In that state, refineries are unable to accept it as feedstock to blend with crude, primarily because of the corrosive action. The primary means for oxygen removal is using hydrogen that is added to produce water. Some of this occurs during the pyrolysis, especially in catalytic pyrolysis. The more common method is fast pyrolysis, in which the constituents are heated at the rate of up to 1000°C per second to between 440°C and 600°C. In this case the bio-crude is more oxygen rich than in catalytic pyrolysis.[3] Catalysts in catalytic pyrolysis are designed to aid the oxygen removal, whereas also preventing charring of the carbon. Minimizing char deposition on the catalyst surface is an important goal. The best of such processes can get the oxygen content down close to 10%, at which point the intermediate becomes tolerable as a refinery feed constituent when blended with crude. Such methods almost certainly add hydrogen to the process. Despite such addition, as shown in Table 11.1, bio-oil is still deficient in hydrogen when compared with conventional oil. This deficit is easily corrected in refining operations, provided the other conditions such as oxygen content are met. Consequently, this intermediate is best directed to an existing refinery, as opposed to setting up dedicated facilities to convert it to transport fuel. Still, if justifiable, a dedicated facility is feasible; the refining steps are the same as in conventional refineries.

Oil seed conversion

Lipids from oil seeds such as rape (canola as a variant) and castor, and tree products such as palm oil, are converted to fuel oil by transesterification.[4] In this batch process, the lipid derived from pressing the oil seeds is reacted with either methanol or ethanol in the presence of a base catalyst. The product is a long-chain mono-alkyl ester that is a drop-in substitute for diesel. The byproduct glycerol has significant impurities such as unreacted alcohol and unused catalyst. So, although glycerol has value, the cleanup could be prohibitive in a small operation. Smallness is a notable point. This is a simple operation and could be done on a farm. The process was refined during the Second World War, where the glycerol was used in explosives. But even if the glycerol is recovered economically, overall economics does not favor this method of biodiesel production. The exception could well be *Jatropha* conversion. The plant grows in arid climates, and cultivation for oil production was first begun in Australia. India and other countries saw major potential. A review in the prestigious journal *Nature* stoked the fires of optimism.[5] But early excitement was dimmed by realities discussed in the following box that nevertheless also offers hope for the future. At the time of this writing, *Jatropha* remains interesting but not relevant at scale.

Jatropha: Boom, bust, and tempered expectations

The promise of *Jatropha* (Fig. 11.2) has been centered largely on two considerations. One is that it bears an oily seed, with as much as 40% oil, depending upon the strain. Unlike corn, soybeans, canola, and other seed plants, *Jatropha* is not also a food source. The second point may, however, be the key. Jatropha can grow in nonarable soil with very little water usage. The fact that it is drought resistant does not mean that it does not grow better with more water. It does. But optimal water consumption is not known and likely varies with strain. Consequently, a farmer with access to water

Fig. 11.2 *Jatropha* fruit on the bush. *(Courtesy: Rulkens T. Jatropha curcas—Fruits. Wikimedia. 2012. https://commons.m.wikimedia.org/wiki/File:Jatropha_curcas_-_fruits_(4729505466).jpg Accessed July 22, 2020.)*

will use it, especially because in most countries, including the United States, water for agriculture is priced very low. In India there is a push to grow it in nonarable areas without irrigation, coincident in many cases with poverty.

Against this backdrop is another problem. All strains grown during the frenzy years of 1999–2010 were wild type; none has been domesticated. Therefore the yields were unpredictable. Uncertain yields make sizing of a processing plant difficult. This is the problem being addressed by a few startups in Singapore, the United States, and Europe. They are all trying to create a strain with predictably high yield and other valuable characteristics such as early flowering, favorable blend of esters (most suited for later processing), and drought resistance. Virtually, all players are breeding new cultivars, many genetically modified to obtain desired properties. In doing so, some of them are using a relatively new and powerful technique known as high throughput screening (HTS). This robotically controlled process allows tens of thousands of experiments to be conducted very rapidly. It also allows one to zero in on the promising subset and quickly perform further optimization on just that subset. The technique has been around for a while and has recently become dramatically cheaper.

HTS has caused an explosion in data generated that needs rapid analysis. Fortunately, this happened coincidentally with new computational schemes to handle the onslaught. The associated field of data analytics is fast growing, and the colloquialism Big Data applies to it in many fields, including drug discovery and rapid diagnosis of genetic defects. In Chapter 14, we will discuss recent startling research using data analytics and demonstrating statistically significant increases in crime in London on high air quality index days.[6]

In 2013, the San Diego startup SGB Inc. claimed significant advances. They conducted HTS-mediated deciphering of strains with known phenotypes (the physical manifestation of a genetic propensity) such as yield and drought tolerance. While wild-type bushes produce only about 6–8 fruits in a cluster, some of SGB's strains produced over 35. The varieties with the best phenotypes were combined to make new strains. These strains were to be produced with standard grafting techniques, so they would not fall in the class of genetically modified crops. The hype was strong enough to merit a December 2013 story in the *New York Times*.[7] Then, the price of crude oil halved in three calendar quarters, starting in mid-2014, making any biodiesel challenging. SGB was restructured to form the company Resolute Genetics and began to target derivatives for markets other than transportation.

A recent review of the status of *Jatropha*[8] describes several derivative markets other than diesel. It identifies nearly half a dozen companies addressing this, including Resolute Genetics. If the industry can survive on other products, the costs may fall to be competitive with crude oil-derived diesel. I (VR) have been predicting since 2014 that the oil price will remain in a band of USD45–65. So far, that has been holding, and even with weakness in shale oil production (the single biggest reason for low oil prices), low-to-moderate prices appear to be here to stay at least another 5 years. But net oil-importing nations with controlled transport fuel prices, such as India, could still be in play for domestic production of *Jatropha*, even at somewhat higher cost.

India has always welcomed *Jatropha* because it is indigenous and because the country is extremely diesel dependent. India burns 584 million barrels of diesel per year, when compared with only 95 million barrels of gasoline. Much of this is in public transport, especially trains. The significance of this statistic is that *Jatropha* is most easily converted to diesel or jet fuel by using transesterification. Interestingly, this can be done economically on a small scale, practically a garage operation. It has considerable appeal for producing fuel in each village cluster for local consumption. Currently, a high fraction of villages have no electricity grid; any meagre electric power is from burning diesel. A viable local fuel has disproportionate significance.

Some processes are uniquely suited to small scale. Photovoltaic solar is one such. We need to embrace these for what they are and resist the temptation to scale up. Sometimes smaller is simply better. In any case, where energy is concerned, small scale and large scale will coexist. One is not necessarily better than the other. Horses for courses. Some horses run some courses better than others. Ask Kentucky Derby winners about the Belmont.

Liquefied natural gas, compressed natural gas, and natural gas as diesel substitutes

Diesel engines can run on natural gas if modified to be spark ignited. Diesel fuel and equivalents are ignited with compression-induced temperature increase. Methane is similar to gasoline in requiring a spark to ignite the fuel-air mixture. The modification is relatively straightforward because the retrofit is on top of the cylinder. Natural gas at normal pressure is not used in transportation vehicles. It is, however, used in stationary applications such as compressors on locations with ready gas access, for example hydraulic fracturing pumps on oil and gas drilling rigs.

Compressed natural gas (CNG) is held between 3000 and 3600 psi to increase the volumetric energy density. It occupies about 1% of the volume occupied by uncompressed natural gas. That is the good news. But the volumetric energy density is 9 MJ/L, when compared with diesel at 34 MJ/L. For the same range, the tank must be over three times larger. To contain the relatively high pressures, the tank must have a cylindrical cross-section. This design aspect essentially prohibits use as a gasoline substitute in passenger cars, where space is at a premium. CNG is appropriate for short-range work in garbage trucks, school buses, delivery vehicles, and the like. For these applications, the other shortcoming, the need for relatively expensive multistage compression stations, is easily met. The school bus is a particularly good candidate because of the frequent stops and idling in the presence of children. Benefit to the children assumes that the emissions are less harmful from CNG than from diesel. On a mass basis of estimating PM, a lot of evidence appears to point that way. However, on a particle count basis, there may not be much difference.[9] These results are similar to those reported in Chapter 10 for gasoline emissions. Gasoline engines certainly produce less soot, the principal nucleating agent for PM, than do diesel engines. But on a particle count basis, not much difference was reported between gasoline and particulate filter-outfitted diesel engines.[10] This tussle between mass-based regulation and potential regulation of particle count is discussed in Chapter 14.

Then again, the toxicity could be different because the diesel-based PM will likely have more organic coating, as methane combustion is not likely to produce volatile organic compounds. Any organic coating will subsequently be from volatile organic compounds in the ambient air. At a school bus stop, proximal to the tailpipe, the relatively fresh emissions can be expected to be more benign from the standpoint of toxicity.[11]

LNG is the preferred diesel alternative for long-haul trucks. It comprises primarily methane with some ethane, chilled to −162°C to form a liquid. The volume occupied is 600 times less than gaseous methane. Whereas this substantially improves the volumetric density, the energy density at 21 MJ/L is still about two-thirds that of diesel. However, the structure of a truck allows the incorporation of fuel cylinders to give range comparable to diesel carriers. Fig. 11.3 illustrates this point.

Note also the cylindrical geometry of the LNG tank, despite the pressure being close to atmospheric pressure. The liquid is kept at −162°C by evaporative cooling produced by bleed off. Because the engine is supplied with gas, not liquid, presumably the cooling effect of gasification is beneficially utilized. If not, the bleed-off would be a source of undesirable methane release to the atmosphere. Filling up would be done at stations with LNG storage. Transporting the LNG to those locations will be in trucks from the LNG facilities that are mostly on the Gulf of Mexico coast. The gas bleed-off could be considerable, adding both cost and material greenhouse gas emissions. The industry response to this has been the development of small-scale liquefaction units. Originally conceived by General Electric and offered in 2013 as "LNG In A Box," the capability is now available from

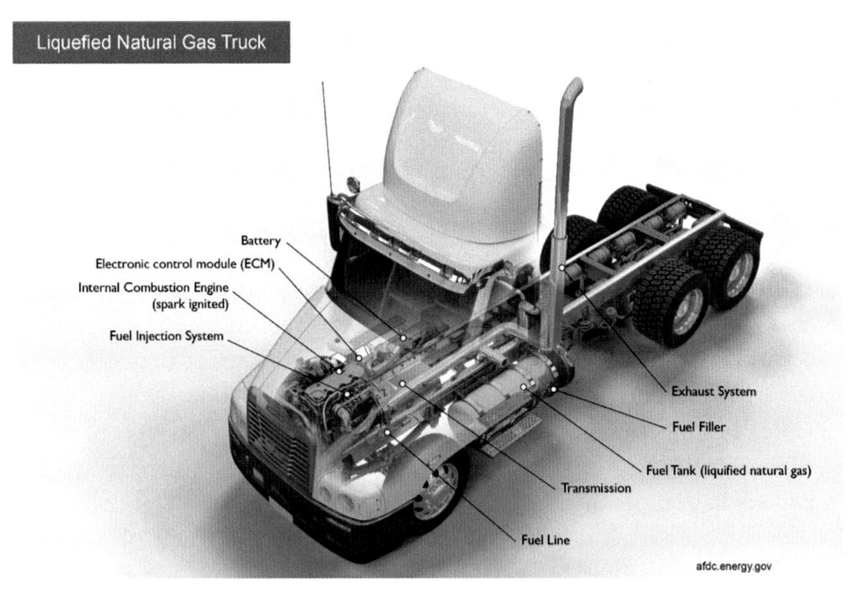

Fig. 11.3 Schematic of a tractor truck fueled by LNG. *(Courtesy: United States Department of Energy. Alternative Fuels Data Center: How Do Liquefied Natural Gas Trucks Work? Accessed July 22, 2020. https://afdc.energy.gov/vehicles/how-do-lng-cars-work.)*

several outfits. Generally, the plants are modular, with capacity ranging from 25 gal to 1200 gal per day (when compared with 9 million gallons per day in a typical conventional plant). In 2018, the market served was already close to USD30 billion. It is expected to increase at a 6.8% compound annual rate over the next decade. This is being driven by demand in the marine and heavy transport sectors and by the low and relatively stable cost of natural gas (when compared with oil) as a raw material. This stability is particularly the case in the United States, where the price has been stable and in the vicinity of a sixth of the price of oil on an equivalency basis. Even after costly liquefaction, it is less than half the price of oil. As with any other distributed production technology, circumstances will dictate whether it is preferred over the conventional large plant and transportation mode.

Dimethyl ether as diesel substitute

DME has the formula CH_3-O-CH_3, with the bonding of the two methyl groups being through the oxygen atom. No carbon-to-carbon bonding means no soot formation on combustion. The long-chain carbon-bonded diesel produces more soot that the shorter chain gasoline. DME has none, so that source of PM is removed. This is the principal allure of DME as a diesel additive or substitute (Table 11.2). As an additive, up to 20% can comfortably be blended with no adjustment required to the engine operating conditions.

The latent heat of vaporization of DME is 467 kJ/kg, when compared with 300 kJ/kg for diesel. Evaporative cooling upon injection into the cylinder allows the engine to run cooler, which results in lower PM and oxides of nitrogen (NO_x) emissions.[12] The higher oxygen content also should allow a more complete burn of the fuel. The substantially higher cetane number indicates better ignition characteristics. The lower liquid density and lower heating value will cause the fuel tank to be larger than for diesel of the same range. DME tanks look roughly the same as LNG

Table 11.2 Properties of dimethyl ether and diesel.

	DME	Diesel
Heating value (MJ/kg)	27.6	42.5
Cetane number	>55	40–50
Liquid density (kg/m^3)	667	831
Enthalpy of vaporization (kJ/kg)	467	300
Oxygen content (mass%)	34.8	0

Data from Arcoumanis C, Bae C, Crookes R, Kinoshita E. The potential of di-methyl ether (DME) as an alternative fuel for compression-ignition engines: a review. *Fuel*. 2008;87:1014–1030.

tanks in being cylindrical and placed at about the same location (see Fig. 11.2). But they are thin walled, and the liquid is stored at 75 psi. The storage and transport of DME are in virtually the same equipment as that for propane. Similarly, the filling is at a relatively low pressure, requiring single stage compression. From the standpoint of combustion, it is more like ethanol, being an isomer. Isomers have the same molecular weight but different structures. Accordingly, the burn characteristics, such as stoichiometric oxygen ratio, are the same as for ethanol that is routinely blended with gasoline and not diesel. Also, the method of manufacture is quite different, as described later in this chapter.

The Oak Ridge National Laboratory conducted a comprehensive study of two DME-fueled Volvo trucks and a diesel-fueled one.[12] Their findings are in accord with the general arguments presented above. They performed steady-state cruising at 60 mph and measured several emissions species. The NO_x figures are shown in Fig. 11.4.

The authors state that all values are below the Euro V standard but note that these were not standard tests. The DME trucks had no NO_x traps, whereas the diesel one had urea-selective catalytic reduction exhaust gas treatment, as described in Chapter 10. The likely takeaway is that DME-fueled trucks without NO_x traps are still compliant with Euro V standards. Also, NO_x and PM tend to be traded against each other in diesel engines, as discussed in Chapter 10. For example, higher temperatures associated with

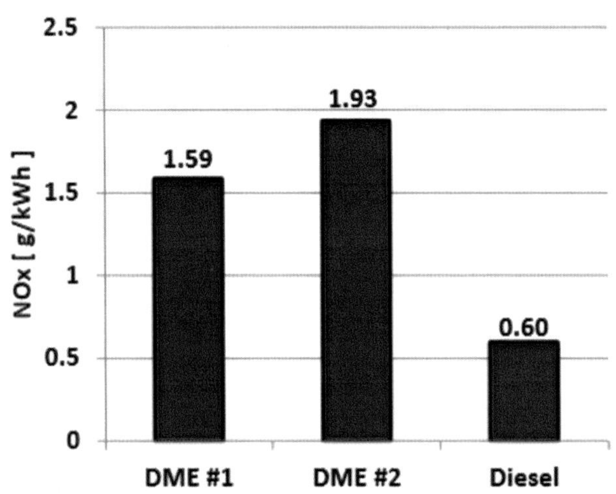

Fig. 11.4 NO_x emissions from DME- and diesel-fueled trucks cruising at 60 mph.[12] *(Courtesy: United States Department of Energy, Oak Ridge National Laboratories.)*

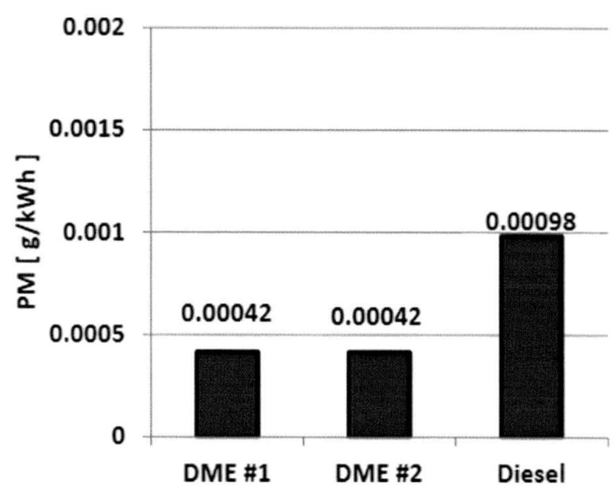

Fig. 11.5 Particulate emissions from DME- and diesel-fueled trucks cruising at 60 mph.[12] *(Courtesy: United States Department of Energy, Oak Ridge National Laboratories.)*

lean burns produce more NO_x, but the higher temperatures in the exhaust are important for regenerating the diesel particulate filter. However, as shown in Fig. 11.5, PM runs so low in DME engines that the engines could be tuned to lower NO_x without concern for PM emissions.

The diesel truck was equipped with a particulate filter. The DME trucks had no aftertreatment but had emissions half that of the diesel. The absence of carbon-to-carbon bonding would have predicted these low levels. These tests appear to confirm that neither NO_x nor PM aftertreatment is necessary for DME-fueled vehicles. The only caution associated with the results is that the PM measurement was on a mass basis. Particulate count measurement may not allow us to be sanguine. But, because DME combustion will not produce aromatics, the particles at the tailpipe may be less toxic than the same number and size from diesel combustion.

Volvo introduced a 13-L DME truck engine in 2013 in Sweden. They and the Mack Truck subsidiary expanded this to the United States in 2014. Then the bottom fell out of the oil (and diesel) price in late 2014. But activity continues, albeit not at the pace that higher diesel prices would have allowed.

Manufacturing DME

DME can be synthesized from syngas, that as noted earlier in this chapter, can be produced through thermochemical treatment of carbonaceous material. In the United States, the commonest such material is natural gas. In India and China, it would most likely be coal. In communities wanting renewable energy and/or local sourcing, it could be biogas from animal waste or landfills. This recounting underlines the versatility of syngas and the materials from which it can be sourced. Syngas is catalytically reacted to convert to methanol. Eq (11.3) is the net reaction. Intermediate reactions involving the formation of CO_2 occur but are not relevant for the discussion.

$$CO + 2H_2 \rightarrow CH_3OH \tag{11.3}$$

$$2CH_3OH \rightarrow CH_3OCH_3 + H_2O \tag{11.4}$$

Catalytic dehydration of methanol produces DME and water (Eq. 11.4). This is the simplest route, provided methanol is available. Oberon Fuels makes small-scale reactors to produce DME at locations close to the filling stations for DME trucks. The water is hydrocarbon contaminated and must be disposed of. The methanol was originally synthesized from syngas, and this is the route of choice. Many have attempted direct DME production from syngas. Although feasible, the process control is tricky, and in the opinion of this author (VR), a methanol intermediate is preferred.

Distributed production of syngas

Syngas is a precursor for most of the fuel alternatives, such as methanol, DME, synthesized gasoline, diesel, and even ethanol, although the last is usually produced biochemically. Most commonly, these fluids are produced in refinery scale plants. A typical methanol plant is sized at 5000 tonnes—40,000 barrels—per day. Synthetic diesel plants employing Fischer-Tropsch's synthesis are usually double that size. Syngas generators are part of the plant. The business model is to conform to these economies of scale and distribute the product in pipelines and tanker trucks, especially the latter for the last mile. The feed to the plants is usually pipeline-conveyed natural gas but is increasingly biogas, especially in states encouraging it. Even when biogas is used, it is often a component, blended with natural gas.

Filling stations located across countries would benefit from local production. If natural gas or biogas is available locally, the alternative fuel could be produced locally. Industry is making a concerted effort through innovation in process intensification to enable small-scale production at or near full-scale unit costs. Startups are targeting methanol, DME, diesel, and gasoline, all synthesized from syngas. The key is economical small-scale syngas generation because each fuel needs this precursor. At this writing, there is only one such at late stages of commercialization. This is the RTI International's MicroReformer™ that economically produces as small a quantity as 10 barrels of resultant methanol per day.[1] The units are modular and can be multiplied to produce more.

Methanol and ethanol in gasoline

Both alcohols are blended with gasoline, and to a lesser extent with diesel. Ethanol is more common and was used as an octane enhancer in the early part of the previous century. Recently, it has enjoyed a resurgence, primarily as an oxygenate (assuring a more complete burn of the fuel) and octane enhancer.

The octane number is a characteristic that defines the ability of a fuel to tolerate high compression engines. In such an engine, the fuel is compressed in the cylinder to a greater extent than in regular engines. Consequently, when it is ignited, the energy released is greater than in the conventional cylinder. This provides high torque to the wheels. More importantly to the fuel economics and the environment, more energy is produced for the same amount of fuel. In simple terms, the efficiency of the engine is increased.

The octane number of the fuel determines whether this can be accomplished. If the octane number is too low, the fuel will ignite prematurely. This is known as "knocking" in the parlance because of the sound produced, a bit like a rattle. Each car engine manual defines the octane rating permitted. Regular engines use 87 octane and most others use 91. Some sporty cars require 93. Therefore gas stations carry three grades. Inexplicably, although, the midgrade is 89 octane, not 91. No car is rated for that value. Ultra-high compression engines such as those in Indy race cars require octane numbers over 100. That is why they use pure ethanol or methanol, both with octane over 110.

But for a given car, higher octane is not necessarily better. The 93 octane grade often carries the descriptor "super" or "hi-test." This can be deceiving. A car designed to use 87 octane gasoline will derive no benefit from the higher grade. This is a case of more not being better.

But more can be better if we change the engine. While this may seem an impractical suggestion, consider first an important fact. Each of the three most viable substitutes for oil-derived transportation fuel has an extremely high octane number. Ethanol, methanol, and methane (the principal constituent of natural gas) clock in at 113, 117, and 125, respectively. Comparing just the liquids in that list, regular gasoline scores an anemic 87.

But ethanol has 33% less energy content and methanol has about 45% less. This explains why E85, which contains 85% ethanol, has never been popular with consumers. Flex fuel vehicles tolerate any mixture ranging from pure gasoline to E85. But without subsidies, E85 costs as much as or more than gasoline, and it delivers 28% fewer miles to the gallon. Not a formula for consumer acceptance.

Methanol is much cheaper to produce than ethanol, especially with cheap shale gas. So even taking into consideration the lower energy content, an M85 blend would be of good value. But one would have to fill up about twice as often as with gasoline.

An elegant solution would be a super-high compression engine, with the ratio around 16. Regular engines are just under nine. At high compression, all three gasoline substitutes would deliver very high efficiency. One could expect the energy density disadvantage to be eliminated. Methanol in particular would be significantly cheaper than gasoline per mile driven. This is more likely with direct injection, where the fuel is injected directly into the cylinder, and one could rely on the high latent heats of evaporation of methanol and ethanol (respectively 3.7 and 2.66 times that of gasoline) to cool the chamber. If done with multiple injections in the compression stroke, the cooling would allow extremely high compression (remembering that compression is primarily limited by the frictional heat causing premature ignition) and allow engine size reduction for the same output as larger engines. MIT professors Daniel Cohn and Leslie Bromberg have been researching this approach since 2006.[13, 14] Recently, they have suggested a gasoline engine assisted with alcohol to rival a diesel engine twice its size.[15] The science certainly holds up. The Mazda SkyActiv engine is likely using this principle in a more modest way.

References

1. Browne JB. *A Techno-Economic & Environmental Analysis of a Novel Technology Utilizing an Internal Combustion Engine as a Compact, Inexpensive Micro-Reformer for a Distributed Gas-to-Liquids System*. Thesis for PhD, New York: Columbia University; 2016.
2. Dayton DC, Foust DT. *Analytical Methods for Biomass Characterization and Conversion. Emerging Issues in Analytical Chemistry*. vol. 10. Cambridge, MA: Elsevier and RTI Press; 2020.
3. Wright MM, Satrio JA, Brown RC, Daugaard DE, Hsu DD. *Techno-Economic Analysis of Biomass Fast Pyrolysis to Transportation Fuels*. Technical Report NREL/TP-6A20-46586, National Renewable Energy Laboratory; 2010.
4. Ma F, Hanna MA. Biodiesel production: a review. *Bioresour Technol*. 1999;70:1–15.
5. Fairless D. Biofuel: the little shrub that could–maybe. *Nature*. 2007;449:652–655. https://doi.org/10.1038/449652a. https://www.nature.com/news/2007/071010/full/449652a.html. Retrieved July 28, 2020.

6. Bondy M, Roth S, Sager L. *Crime is in the Air: The Contemporaneous Relationship Between Air Pollution and Crime*. IZA Institute of Labor Economics; 2018. Discussion paper series, IZA DP No. 11492, April.
7. Woody T. *Start-Up Uses Plant Seeds for a Biofuel*. The New York Times; 2013. December 24, Accessed July 22, 2020 https://www.nytimes.com/2013/12/25/business/energy-environment/start-up-makes-gains-turning-jatropha-bush-into-biofuel.html.
8. Hawkins D, Wigglesworth T. *Jatropha Sector Review*; 2017. https://www.hardmanandco.com/wp-content/uploads/2018/09/too-good-to-burn-the-industrialisation-of-jatropha-1.pdf.
9. Vermeulen RJ, van Gijlswijk RN, Heesen D. *Dutch in-Service Emissions Testing Programme 2015–2018 for Heavy-Duty Vehicles: Status Quo Euro VI NOx Emission*; 2019. https://publications.tno.nl › publication › TNO-2019-R10193.
10. Mohr M, Forss A-M, Lehmann U. Particle emissions from diesel passenger cars equipped with a particle trap in comparison to other technologies. *Environ Sci Technol*. 2006;40 (7):2375–2383. https://doi.org/10.1021/es051440z.
11. Lichtveld K, Ebersviller SM, Sexton KG, Vizuete W, Jaspers I, Jeffries HE. In vitro exposures in diesel exhaust atmospheres: resuspension of PM from filters versus direct deposition of PM from air. *Environ Sci Technol*. 2012;46(16):9062–9070. https://doi.org/10.1021/es301431s.
12. Szybist JP, McLaughlin IS. *Emissions and Performance Benchmarking of a Prototype Dimethyl Ether-Fueled Heavy-Duty Truck*. Oak Ridge National Laboratory; 2014. Report ORNL/TM-2014/59.
13. Bromberg L, Cohn JR. *Effective Octane and Efficiency Advantages of Direct Injection Alcohol Engines*. MIT Laboratory for Energy and the Environment; 2008. Report LFEE 2008-01 RP.
14. Cohn JR, Bromberg L, Heywood J. *Fuel Management System for Variable Ethanol Octane Enhancement of Gasoline Engines*; 2008. US Patent 7,314,033, 2008.
15. Stauffer NW. *Getting the World Off Dirty Diesels*. MIT News; 2018. http://news.mit.edu/2018/mit-researchers-gas-alcohol-engine-getting-world-off-dirty-diesels-0613.

CHAPTER TWELVE

Electric vehicles: Transformational solution for low-PM transportation

Vehicles powered by electricity are the future. The early motivation was displacement of fossil fuels. This displacement was justified largely on the premise that fossil fuel production and combustion had adverse effects on the environment. That justification remains, but a subset is the production of particulate matter, with all the deleterious effects on the human condition that have been described in earlier chapters. This subset may now be the largest driver. Chapter 10 presents evidence for the belief that gasoline-powered cars have been getting something of a free pass in comparison with diesel vehicles. With increased public awareness of this, with or without associated policy changes, the justification for adoption of electric vehicles (EVs) will rise. Motivations such as this may be necessary because EVs have significant hurdles to adoption, and a big one is societal acceptance. These hurdles are being progressively traversed.

History of electric vehicles

The focus here will be primarily on passenger vehicles such as cars and light trucks. Other applications such as trains will be mentioned but not dwelled upon, in part because the hurdles are different and the timelines for exploitation are different. For example, electric trams never had a hiatus in usage as did electric cars, in part because public sector undertakings with high initial capital tend to have different drivers. Furthermore, public transport engines are more amenable to overcoming some of the hurdles, such as availability of electric power.

Electric cars experienced a hiatus of nearly a century in being a material portion of the fleet on the road. They were the first commercially available and widely used cars, mostly in Great Britain and France, c.1890. Dates are hard to pinpoint because of the variants. Initial vehicles used primary cells (not rechargeable) in their batteries, which presented economic and

Particulates Matter
https://doi.org/10.1016/B978-0-12-816904-9.00024-6

logistical challenges. Vehicles became available after the invention of rechargeable batteries using lead-acid chemistry in the 1850s, but it took electric storage capacity improvements (to provide acceptable range on each charge) until the 1880s to become commercially useful. Englishman Thomas Parker, who was also responsible for the electrified London Underground railway, is credited with designing and building the first commercially viable electric car. Part of his motivation is believed to have been to address smoke and other combustion emissions in London at that time.

Gasoline-powered cars did exist in that era, but EVs had several features that proved attractive. They were not noisy and smelly. They did not require gear changes. Before the invention of the synchromesh in 1919, offered commercially a decade later, changing gears was tricky because it involved making the change at precise revolutions per minute; ineptitude would cause a mismatch of the spinning gear teeth, resulting in a grinding sound and potential damage to the teeth. But the key attribute may well have been the ability to switch on the engine without manual cranking. For readers not familiar with cranking, it involved inserting a rod with an offset handle used to turn it. The front end of the rod had an "ear" on diametrically opposed sides, which fit into a receptacle in the engine. Turning the rod clockwise, using the offset handle for leverage, caused the engine to fire. Women (especially posh women, who were the only ones who could afford the early vehicles that were ornate) were taken with electrics. This predilection is believed responsible for the allusion to EVs as "women's cars," not unlike the modern characterization of soppy romantic films (think *Notting Hill* and *Sleepless in Seattle*) as "chick flicks." Even the early Mazda Miatas were credited with this associative allusion.

The proliferation of EVs was hampered by the limited range of 40 miles or less (interestingly, that was almost exactly the electric range of the hybrid Chevrolet Volt when first introduced) and low top speed (presumably because higher speeds discharge batteries too fast) of 20 mph. The range was a constraint because of limited recharging infrastructure (sound familiar?). How the industry attempted to solve this (with battery swapping) is described later in this chapter. The same means is being used to address the current hurdle, yet another illustration of how the more things change, the more they stay the same (*plus ça change, plus c'est la même chose*; Jean-Baptiste Karr 1849).[1]

The time in the sun for EVs lasted a scant two decades before the eventual shade set in. It all began when, on January 10, 1901, an oil well on what came to be known as the Spindletop field blew out, producing 100,000 barrels each day for 9 days before it was brought under control.[2] This was the start of the Texas oil boom and the age of oil. Geologists had believed that the rock structure would contain oil, so this was not a surprise, although the size of the field might have been.

Prior to Spindletop, as it is colloquially referred to as a historic marker in the annals of oil, the only uses of oil had been for lubricants and as lamp oil. Continued development caused the price of oil to plummet from USD 2.0 to 25 cents a barrel, and it became feasible for use as a combustible fuel. This helped usher in the internal combustion engine (ICE) as a mainstream transport means. Relatively cheap gasoline-driven engines were more economical and had much greater range than EVs. The previously invented muffler on the engine exhaust was ready for implementation to address the noisiness. Even the pesky engine crank went away when Charles Kettering invented the electric starter in 1912. On this point, I am forced to make a personal note. As a boy of six in 1950, I (V.R.) was still crank-starting the family Morris sedan. Inventions do not translate into widespread use quickly, especially in a secondary market such as India (at the time). Furthermore, legacy fleets stay on the road a long time; an instructive lesson for those wanting EVs to be dominant any time soon. The final nail in the EV coffin was Henry Ford's invention of assembly-line production, which further drove down the price of gasoline-powered cars. The famous 1909 statement by Ford, "Any customer can have a car painted any color that he wants so long as it is black," while ill-advised from a marketing standpoint, was driven solely by the desire for simplicity to keep the Model T affordable. In that, he succeeded.

Hybrid electric vehicles

The first modern-day production EVs were not all-electric, as had been the historical precursors. They were hybrids of electric power and ICE power. The first was the Toyota Prius, introduced in Japan in 1997 and in the United States 2 years later. It was a full hybrid, in that the drivetrain could accept electric propulsion alone or ICE propulsion or both. Later, lower-cost hybrids had serial architecture, where the drive is solely electric and a small ICE charges the batteries.

Fig. 12.1 shows the features of a plug-in version. Note that all aspects of an ICE are present and that the additional elements are a battery pack and an electric traction motor driving the front wheels.

The Prius introduced key new features such as regenerative braking. This is an old invention that uses the energy lost in braking to charge the batteries.

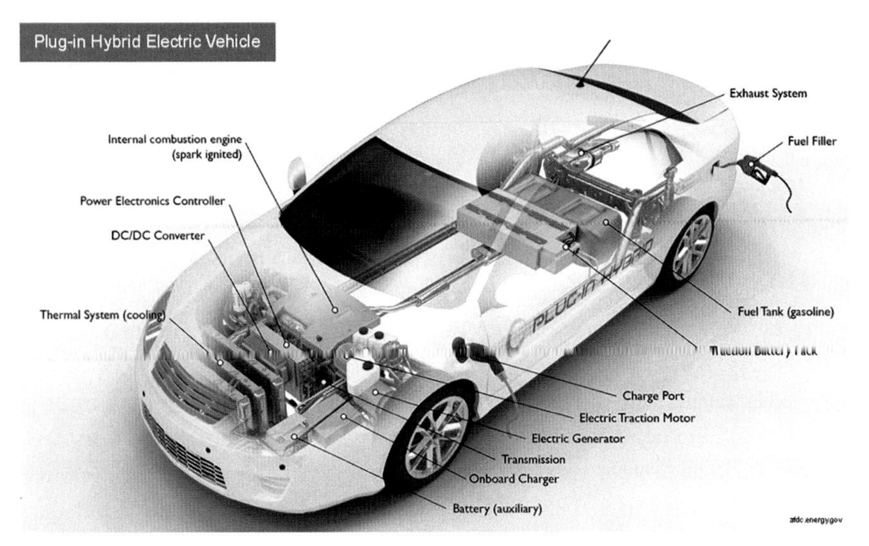

Fig. 12.1 Plug-in hybrid electric vehicle. *(Courtesy: United States Department of Energy.)*

Another feature was engine shutoff during idle. Electric motors are much more reliable for stop-and-start than ICEs (although replacing carburation with fuel injection helped a lot), thus safely enabling this feature. These cars are placed in electric drive at the start and up to a certain speed, when they switch to the ICE. In combination, these two features improved fuel efficiency so much that miles per gallon (mpg) figures were higher in city driving than on the highway. The sealed battery pack comprised nickel-metal hydride cells.

Each pack had up to 15 kg lanthanum and 1 kg neodymium, both from the rare earth family of elements. This application is credited with creating the fear of scarcity of rare earths, most of the world's supply coming from China, the low-cost producer. The constraint is not the total reserves of these elements. It is that most other producers are unable to meet Chinese prices (if not cost to produce).

These battery packs are a fraction of the size of those in all-electrics. Even the second-generation Prius pack had 1.31 kWh capacity. Only a portion of this capacity was in use, to have shallow discharge (discharging down to nearly zero capacity shortens battery life). As a rule of thumb, a mile driven consumes 0.25 kWh. If only 1 kWh of the capacity is in routine use (if push came to shove, literally if the ICE died, electric drive could use up the entire capacity), this translates into a range of 4 miles. This is ample for the primary purpose of electric drive after a stop and until a moderate speed is achieved.

Also, regenerative charging will keep topping up the pack, in effect increasing the electric range for a given journey. Hybrids have the unique feature of using electric drive in revolutions per minute regimes where ICEs are inefficient. This aspect and zero idling are likely the main ingredients for the vastly improved miles per gallon over a conventional engine.

The early hybrids, in 2001, were rated at 40 mpg in city driving. This increased to about 52 mpg by 2010. At that figure, they were about twice as fuel efficient as the same vehicle run only on the ICE. This addressed the emissions objective by simply driving more miles per gallon of gasoline. An all-electric has zero tailpipe emissions. Electricity production produces emissions if fossil fuel is used, but large plants are more amenable to cleanup of combustion products than are tailpipes on cars. This question of "how dirty is your electricity?" is addressed later in the chapter.

The next evolution was the plug-in hybrid electric (PHEV). The early embodiments merely topped up the small nickel-metal hydride pack, which could be done relatively quickly. When General Motors introduced the Volt, it compromised using a large lithium-ion battery pack with 40-mile range, and an ICE with a full gasoline tank extended that up to 500 miles. Charging time was significant, but in many ways the Volt eased the public into the electric car world gently.

Plug-in hybrid electrics are electric vehicles on training wheels

Training wheels are a wonderful invention to aid the tot with two-wheel transport anxiety. More often than not, the anxiety resides with the parents, but regardless of source, the wheels get installed. Now, in purely engineering terms, the extra wheels are pedestrian in design. Clearly intended for the short term, they are not of particularly robust construction because not much use is anticipated. The added cost is modest when compared with that of the bicycle. Yet, the comfort to the psyche is enormous. All this really only applies to the munchkins. Were you to learn to ride a two-wheeler at an advanced age, as was I (V.R.) at age 11, the training wheel option is essentially out. Even if available for a larger bicycle, the derision of the cohort group would not be sustainable. So, what does all of this have to do with electric cars?

Electric cars come in two flavors: all-electric vehicles and PHEVs, both with the ability to conveniently plug into wall outlets and both utilizing the energy of braking to charge a battery. Both can use electricity alone to drive the wheels, so there is an essential simplicity to the mechanics: no transmission, no gearbox, no camshafts, and minimal mechanical maintenance. The principal difference between the two is the auxiliary gasoline engine in the hybrid electric, which in some variants will charge the battery if it runs down. In either case, the ICE offers range beyond the electric. The all-electric does not have this backup feature. So, it will rely solely on the battery for range. The early-entry vehicles had an electric range of 40 miles for PHEVs and 80–100 miles for EVs. These figures more than doubled by 2018, with a few expensive ones up in the 300-mile range.

The car-buying public will face a choice. Because the mass-produced all-electric vehicle is expected to be cheaper to make despite the bigger battery, the list price will be lower than that of a PHEV, with one manufacturer projected to offer it at a price comparable to the gasoline counterpart. The PHEV, on the other hand, although more expensive, will have the much greater range afforded by the gasoline backup. The "fuel" costs will be comparable when run on electricity. The key difference is a new term that has entered the transport lexicon: range anxiety. We can roughly define this as the fear of running out of juice without a convenient fill-up station. The buying public will have some fraction afflicted with range anxiety. This is where PHEVs play the role of training wheels. With such a vehicle, consumers have the luxury of sorting out their driving habits, their discipline in charging every night, and all other manner of behavior impinging upon their ability to live with the limited range of an all-electric vehicle, at all times secure in the notion that the gasoline engine can bail them out. There will also be a segment of the population eschewing this aid to behavior modification, in effect wobbling on to the bike, as yours truly did some decades ago. A skirmish with a thorny bush sticks, as it were, in the memory. Thorny situations will undoubtedly lie in wait for first-time EVers. And then again, perhaps PHEVs will always have a place. Choice is a good thing, in cars, colas, and presidential elections.

All-electric vehicles

All-electric vehicles, also known as battery EVs to distinguish them from hydrogen fuel cell vehicles (FCVs), rely entirely on batteries for propulsion. Hybrid vehicles use ICEs as the primary mode of propulsion. Total reliance on batteries introduces constraints that are discussed below. Fig. 12.2 shows such a vehicle. This is a front-wheel-drive version with a single electric motor and a transmission line to the wheels. Some versions have a motor for each wheel. The battery bank is on the floor in the middle. It is a heavy component and accordingly is set low and centered in the location with the greatest stability. This location also allows ease of replacement from below. Note the simplicity and clean lines when compared with gasoline engine cars.

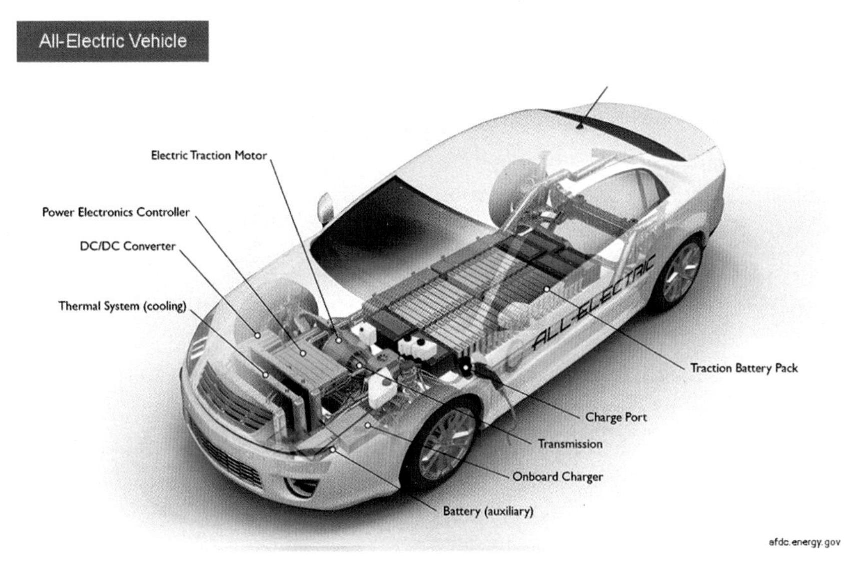

Fig. 12.2 Battery-driven electric vehicle. *(Courtesy: United States Department of Energy.)*

Batteries

In the 1967 film *The Graduate*, Dustin Hoffman's character was given unsolicited advice regarding the career direction of choice: "plastics." At the time, this was either misinformed or prescient. Today, one word unmistakably is deterministic of the future of EVs: batteries. In fact, this has been the case from the very beginnings of EVs over one and a half centuries ago. Except for public transport such as trams and trains, which could afford the capital cost of permanent emplacements of power lines, electric transportation has always faced the problem of stored electricity.

Batteries come in two broad classifications: primary and secondary. All batteries comprise connected portions known as cells. A battery pack has a combination of cells, with the connective elements adding a layer of cost. Later, when we discuss the cost of batteries, careful attention must be placed on the cost per cell versus per battery. Primary cells are a single-use item. The best commercial example is the batteries in portable devices such as powered headphones. The most common primary battery contains alkaline cells. They are cheap, long-lasting in storage, and fit for purpose.

The earliest and still the most robust secondary battery is the lead-acid type. Invented around the middle of the 19th century, it transformed many applications. It remains the sole electric power source for ICEs. The longevity is premised in part upon the robustness. It ought to last for well over

100,000 miles of usage and to withstand several deep discharges (a "dead" battery that you charge up by giving it a jump from another vehicle and then driving around till it recharges; we have all been there).

The lithium-ion battery can be considered the single greatest enabler of EVs. In the discharge cycle, lithium ions move from the anode (negative electrode) to the cathode through an electrolyte. Ion transport reverses in the charging cycle. Most of the advances have come in the materials for each of these components, in particular the cathode.

It was invented c.1979, principally by John Goodenough, Akira Yoshino, and Stanley Whittingham. They shared the 2019 Nobel Prize for chemistry for this discovery, and the citation included the statement "laid the foundation of a wireless, fossil fuel-free society."[3] Whittingham proposed and built the first rechargeable version, using a lithium metal anode and a titanium disulfide cathode. This was expensive and had safety concerns. Goodenough improved it with a layered oxide cathode impregnated with lithium ions. The energy density was vastly improved. To this day, lithium cobalt oxide is the cathode of choice, making cobalt availability an issue.

Yoshino's contribution includes manufacture in 1991 of the first production-viable and safe lithium ion battery at the Asahi Kasei Corporation in Japan. He invented carbon-rich anodes impregnated with lithium ions. The safety issue has dogged the lithium-ion battery, with explosions in portable computers and cellphones often reported. This notwithstanding, electric car batteries have not had this issue, likely because it was addressed at the outset in the control architecture.

The energy density per unit weight of the lithium-ion cells is five times that of the venerable lead-acid. Lead is heavy, with a specific gravity of 11.3, that is the reason for use in radiation shielding. The only denser commercial material is depleted uranium. This is U_{238} after substantially all the fissionable U_{235} is removed, hence the term depleted. Early Boeing 747s are believed to have used over 500 kg per aircraft for counterweight applications. But the high density makes the lead acid cell heavy when compared with both nickel-metal hydride and lithium-ion. Lithium, on the other hand, is the lightest metal, with half the specific gravity of water, at 0.53 g/cc. However, it is extremely flammable, especially in powder form, the likely reason the first commercial batteries went away from lithium metal anodes.

Lithium-ion cells have another advantage: higher working voltage, almost double that of lead-acid and almost three times that of nickel-metal hydride. Fewer cells are required to achieve a needed voltage. The Coulombic efficiency (discharge/charge) is also highest, at 99%, compared to 90%

for lead-acid and 70% for nickel-metal hydride. The cycle life, an important metric for rechargeable batteries, of about 500 cycles, compares well with 400 for nickel-metal hydride and 300 for lead-acid. Lithium-ion will most likely continue to be the workhorse for EVs and portable electronics. While other chemistries are being researched, they will probably be relegated to stationary applications such as backup storage for renewables because low weight will be the dominant attribute in mobile applications.

Economics of electric vehicle batteries

The single largest hurdle for widespread adoption of EVs is battery cost. Here the reference is to capital cost; operating costs are discussed separately. In recognition of this, manufacturers have attacked the problem and made astonishing progress in the last decade (Fig. 12.3).

For an invention that was a scant 20 years old at the start of this graph, the slope of cost reduction is startling. Admittedly, a high fraction—likely over—30% of the battery cost is administrative overhead, and mass production will reduce this, at least as a proportion. Another factor could be that reduction in the price of lithium carbonate (the commodity from which most lithium-based components are made) has been driven by competition. Most of it comes from South America, and the economically depressed Bolivia became a large player. The South American nations, including Chile

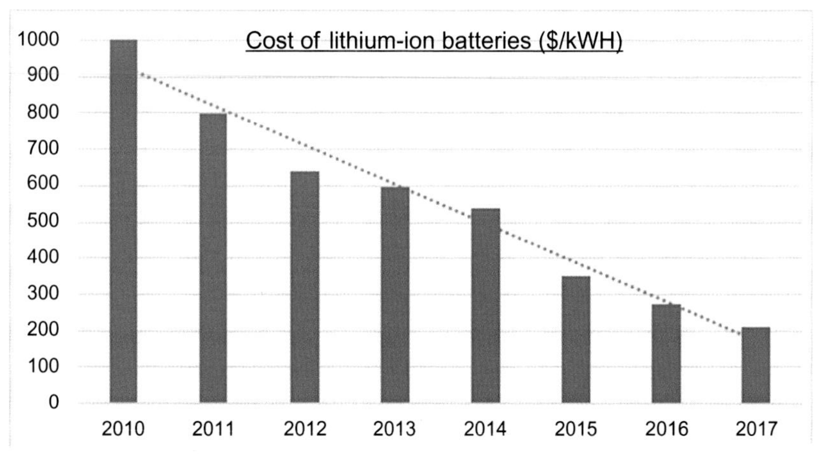

Fig. 12.3 Lithium-ion battery cost reduction trend. *(Adapted from A Behind the Scenes Take on Lithium-Ion Battery Prices. Bloomberg New Energy Finance. https://about.bnef.com/blog/behind-scenes-take-lithium-ion-battery-prices/; retrieved March 5, 2019.)*

and Argentina in addition to Bolivia, benefit from the source being lithium chloride deposits; in Bolivia they are known as *salars* and are salt deposits with high concentrations of lithium. Mining consists simply of pumping down water and recovering concentrated brines; very cost-effective. In contrast, a major alternative source is the ore spodumene with the formula $LiAl(SiO_3)_2$, which must be mined and the Li component separated. This process is more costly than simply producing LiCl from brines. Ultimately, depending on a few countries for lithium sourcing may become untenable (one is reminded of the monopoly in oil created by the Organization of Petroleum Exporting Countries cartel). The solution will lie in recycling the lithium, which is technically feasible. The lead-acid battery industry is enormously helped by the fact that all the lead from the batteries has been recycled for decades now. This made the primary lead production (from ore deposits) passé from the standpoint of both potential scarcity and cost.

A point of note is that the source for Fig. 12.3 uses average cost numbers; the leaders ought to be doing better. Furthermore, companies are secretive about their actual costs. The exceptions have been General Motors and Tesla. General Motors disclosed in 2018 that their battery cost for the new entry all-electric Bolt (great name, by the way, because of the high acceleration feasible from a standing start for all-electrics, which is due to the linear torque/rpm (revolutions per minute) profile) to be USD 145 per kWh, with an expected target of USD 100 by 2022. Interestingly, at the 2018 Tesla shareholder meeting, responding to a question on battery cost, Elon Musk took the microphone away from the more circumspect Chief Technology Officer and stated that USD 100 per battery would be possible by 2020.[4] Tesla makes its own batteries and, as the largest producer, could be expected to have the best economics. Committing to the USD 100 figure that many, including this writer, consider the tipping point for broad-scale use of EVs, was a signal event.

As a rule of thumb, each mile driven uses 0.25 kWh. Some manufacturers have mentioned figures as low as 0.20, and certainly the actual consumption depends on design factors, including vehicle weight. A 100-mile range requires 25 kWh in principle. But it is impractical to drain down to zero, and a useful maximum drain is likely 80% or 85% of capacity. In other words, a 100-mile range needs a battery pack with about 30 kWh. Charging from regenerative braking adds a further cushion. But for purposes of comparison, we will stick with 0.25 kWh per mile. At the 2010 cost of USD 1000 per kWh, the battery pack for a 100-mile range would cost USD 30,000, prohibitively high when compared with a gasoline car with the same other attributes. Manufacturers would have to rely on customer motivations other than economic. In fact, in the early days of the Prius, a PhD thesis at the University of California, Davis, examined the motivation of Prius owners. The graduate student concluded that the primary motivator was "to be seen as green".[5] Early adopters often have unique motivation, but large-scale adoption will be driven dominantly by economics.

At USD 100 per kWh, the pack would cost USD 3000. That would be a very reasonable fraction of a target selling price of USD 30,000. Considered in another way, the figure would be comparable to the cost reduction of not having an ICE, transmission, gearbox, and differential, to name the principal elements of an ICE car that would now be eliminated.

Operational cost would be dominated by the cost of electricity. If most EVs were charged at night, the cost would be lower because most jurisdictions in the United States now have time-of-day pricing, with night time much lower. A figure of USD 0.03 per kWh is reasonable for purposes of computation. This is also a bit higher than the winning bid for a utility scale solar facility in India.

Here is a comparison of a 100-mile range and a 200-mile range EV with a hypothetical ICE car delivering 30 mpg.

EV cost of electricity for 30 miles: $30 \times 0.03 \times 0.25 =$ USD 0.225 or 22.5 cents.

Add an expected highway tax (about 50 cents per US gallon for gasoline): 10 cents.

Amortization of USD 3000 battery over 150,000 miles: 60 cents per mile.

Amortization for battery with a 200-mile range: 120 cents per mile.

Total cost per 30 miles driven: 92.5 cents.

Break-even gasoline cost for a 100-mile range EV: US$0.925.

Break-even gasoline cost for a 200-mile range EV: US$1.525.

Currently, electricity in EVs is not taxed. The figures above are just plug-in numbers to recognize the inevitability of such taxation. Even adding 40 cents more tax per 30 miles (a gallon of gasoline equivalent) does not materially change the comparison because gasoline prices are rarely under USD 2.00 per gallon. In Europe, they are double that.

The takeaway is that the fuel cost break-even with gasoline makes EVs economically viable at battery costs in the vicinity of USD 100 per kWh.

Battery exchange electric vehicles

In this business model, the consumer buys the vehicle without the battery pack. The battery pack is leased, and the consumer pays a variable cost per mile. When the battery pack is near depletion, it is swapped for a fully charged one. The Hartford Electric Light Company offered this service through a battery subsidiary in the period 1910–1924. This timing coincided with the introduction of Ford's Model T, so it might have been intended to reduce the capital cost of the vehicle that was rapidly losing the price battle with the Model T. Ultimately, the exchange business model died with the rest of the EV business.

Fast forward to 2007. A startup, Project Better Place (that later dropped the first word) was launched by the charismatic Israeli Shai Agassi. Agassi's rock star persona and passion attracted nearly USD 1 billion in investment and the support of important players, including Renault or Nissan chief Carlos Ghosn. Renault contracted to make cars for Better Place. The Better Place message was the same as that of Hartford Electric: treat the battery and the electricity consumed as a variable cost, reducing the purchase price of the car. Agassi had an important variant: he would purchase the cars and sell them as Better Place-branded cars. The company would own swap stations at intervals along highways, with a swap time of 5 min, comparable with that of a gasoline station stop. Instead of incurring the inventory and carrying costs of just the batteries and swap equipment, the company would also carry vehicle inventory. It also redesigned the vehicles to include internet-enabled features. It declared bankruptcy in 2013 after selling just 1500 cars. Much has been written about this spectacular rise and flameout. Because of a lot of alleged mismanagement and profligacy,[6] principal causes of failure are hard to discern. But the high cost per kiloWatt hour of the battery had to have been a factor. Interestingly, the Better Place cars had an 80-mile range, close to our hypothetical case. The 30 kWh battery pack would have cost USD 30,000. The car without the battery pack was slated to retail for USD 35,000. The inventory costs for the batteries would have been prohibitive. The lengthy cited piece by Max Chafkin makes no mention of the estimated cost of the battery pack, nor the significance of the inventory costs; possibly it was swamped by the other considerations. Leaving aside the issues that dogged the company, the concept ought to be viable at lower battery prices. That brings us to today.

Battery swapping may have found its place in the sun due to the plummeting cost of packs, as low as US$100 per kWh by 2020. Battery exchange electric vehicles (BEEVs) will be more economical for consumers than the conventional purchase-and-charge vehicles. The business model may also provide a net benefit for other stakeholders, including the likelihood of vastly improved sales that in turn would provide volume for a battery swap business. Stakeholders might be affected as follows.

Auto companies

The BEEV is a very simple machine, squarely in the core competency of auto manufacturers. Batteries, on the other hand, are technologically not in the mainstream, and avoiding the responsibility for anything more than designing the vehicle that houses them would be welcome. Auto companies would be responsible for the control system but that too is familiar from experience with the hybrid vehicle portion of the fleet.

Battery packs may have to be standardized, requiring cooperation or legislation. If the automakers believe that the business model will help sell cars, they will probably cooperate. Tesla, although not drinking much of the Better Place Kool Aid, placed the battery module in an accessible location on the bottom of the Model S sedan and on June 30, 2013 (3 weeks after the Better Place bankruptcy filing) demonstrated a 90-s swap at an event.[6] One design feature could be for the replaceable packs to be in modules, most likely six, with an expectation of four, maybe five, to be swapped per "fill." Packs may advantageously have "history" on board for giving full credit for the residual charge and to understand the severity of the usage. At some point when the swap model matures, car manufacturers could consider removing the expensive charging port from BEEVs. Another option to be considered is a spare backup battery for the truly range anxious.

A battery swap operation would be a substitute for gas stations, where much of the profit is currently in the associated convenience store. The margin could be expected to be higher than that with gasoline. Gas stations today are largely franchised by major oil companies. Battery swap stations would, almost by analogy, be owned or franchised by electric power companies, as was the case with Hartford Electric Light during the first coming of the battery swap model in 1910.

Consumers

Consumers will welcome this business model. The purchase price of a BEEV will be lower by thousands of dollars than that of a comparable car with a nonswap battery. The pay-as-you-go aspect will appeal, especially the differed cost aspect. A car is an asset with a low duty cycle, a high proportion of unused time. With this model, at least the expensive battery portion is not part of that idle asset. Finally, the per-mile fuel cost comparison with gasoline could be communicated simply. The public is used to new cars having stickers with that sort of analysis.

The swap model eliminates the decision of the type of charging equipment in the garage. 240 V systems charge faster but are more costly to install.

It also eliminates the owner deciding the time of day or night to charge the batteries. In short, less nuisance. Many apartment dwellers have no such decisions to make, and the business model appears tailored for them. Even the travel to the swap station is no different from a visit to the gas station to top up. Innovations such as electromagnetic charging (no plug-in required) would fade.

Swap station owners and franchisers

The battery swap operation is purely mechanical, requiring no competency in battery technology or even electric equipment. The equipment will probably be robotic, rather than manual in the Western world. As with the Tesla Model S, the pack will be on the bottom, in part for assuring a low center of gravity. The swap robot would be in a well over which the car parks. The takeaway is that the operation is easily franchised by the electricity company.

The (nominal) 150,000-mile battery pack lifetime would be achieved faster than if owned by the consumer, so the amortization schedule is faster for the franchiser. Modules could advantageously be charged in a hub-and-spoke model; not at the station. This allows economies of scale. Also, it makes solar viable: more space for solar panels in inexpensive real estate, and battery banks could be buffer storage. The electricity could be acquired from a variety of sources. Charging could be done at optimal temperatures and times in designated buildings, rather than in uncertain ambient conditions. Expect the cost per kiloWatt hour of charge to be less than for that at an individual battery recharge station, and expect the batteries to perform better and longer. Finally, advances in batteries can more easily be retrofitted with this model than with batteries in consumer possession.

Swap *versus* fast charging

Fast charging is increasingly the flavor of the day. Our convenience society wants ubiquitous charge stations with minimal wait time. All that equates to fast charging that entails charging in minutes rather than hours. Recent advances in wide-bandgap semiconductors allow chargers to be relatively small and charge times short.[7] However, conventional electrochemical wisdom is that this will bring baggage from reduced battery life and safety. Advances in batteries to allow fast charging will be necessary, and some are being researched.

Swap charging, if conveniently located, will put quite a serious dent in fast-charging stations.

Electric vehicles use less energy

The motivation for adoptions of EVs and FCVs are many. Zero tailpipe emissions, especially in urban settings, is one. Another is the substitution for fossil fuels, although that is tempered by the means for electricity and hydrogen generation, which dominantly use fossil fuels today. That is where the title of this section comes in play. Even if fossil fuels are used to produce the electricity, if the engine uses less energy overall, then the fossil fuel used per mile is less.

EVs expend fewer units of energy than internal combustion vehicles to travel the same distance. This is important, because simply using less energy to receive the same gratification is a powerful arrow in our carbon mitigation quiver.

One means of comparison is the miles driven per gallon equivalent of electric battery charge (MPGe), defined by the United States Environmental Protection Agency using the estimate of the energy in a gallon of gasoline to be 33.7 kWh. The Chevy Bolt clocks in at 119 MPGe.[8] That makes it about three times as fuel efficient as the hypothetical gasoline engine car rated at 35 mpg. This calculation indicates that a Bolt travels a mile with one-third of the energy of a gasoline car of similar size. Put another way, the gasoline engine uses 33.7 kWh for 35 miles driven. This translates into 0.96 kWh per mile. The Bolt uses 33.7 kWh to travel 119 miles, which equates to 0.28 kWh per mile. The takeaway is that *an electric car uses about a third of the energy for a mile driven as does a gasoline car of the same size*. Using less energy to travel a mile, no matter the source of the energy, is a compelling reason for EVs. This squarely addresses the refrain "how clean is your electricity in the EV?"

Hydrogen fuel cell vehicles

California is doubling down on FCVs (Fig. 12.4), despite being home to the world's most prominent EV manufacturer, Tesla. The state is building a network of heavily subsidized filling stations. FCVs are EVs with a different source for the electrons driving the electric motors. In EVs, it is batteries. In FCVs, it is fuel cells. Hydrogen is the fuel, and the output is electricity and water. Both vehicles are zero emission if water, quite properly, does not count as an emission.

Comparing with the all-electric model in Fig. 12.2, one notes that the front end housing the motor and control systems is the same. Electric motors

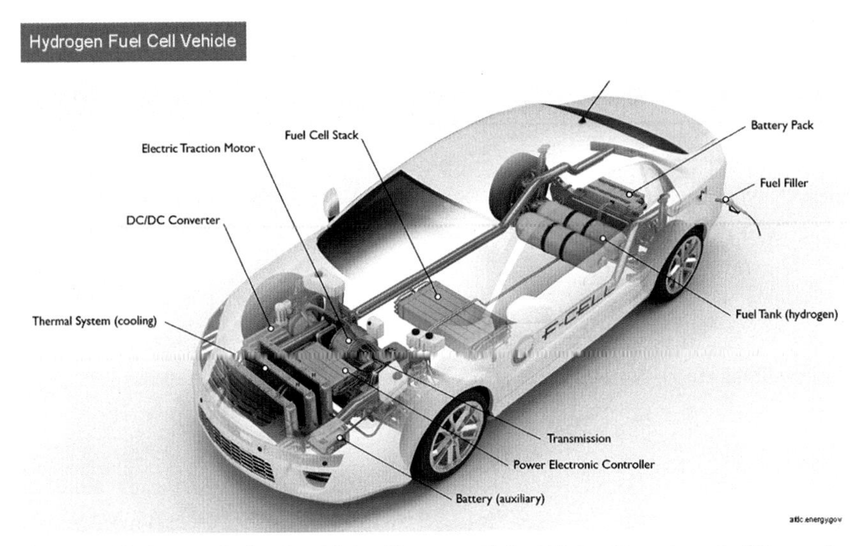

Fig. 12.4 Hydrogen fuel cell vehicle. *(Courtesy: United States Department of Energy.)*

are agnostic about the source of electrons. In principle, legacy EVs on the road could be modified to run on hydrogen. Practically, though, the batteries in EVs have a different form factor and location from the gas tanks and fuel cells. Batteries in the newer vehicles such as the Tesla Model S are located on the floor in the center of the car. The fuel cells in FCVs are typically also on the floor near the center but occupy much less space (Fig. 12.4). Hydrogen gas tanks are in the back. Also, the gas tanks are cylindrical in profile, geometrically a very inefficient form, not unlike natural gas vehicle cylinders. This is due to the storage pressure; when it is high, safe storage can only be in a cylinder. For natural gas, the pressure is 3600 psi, and even that needs cylindrical geometry. The FCV tanks are at 5000 psi or optionally 10,000 psi. In addition to the inconvenient geometry, the high pressure requires multistage compression at the filling station, which increases the cost of the station (Fig. 12.5).

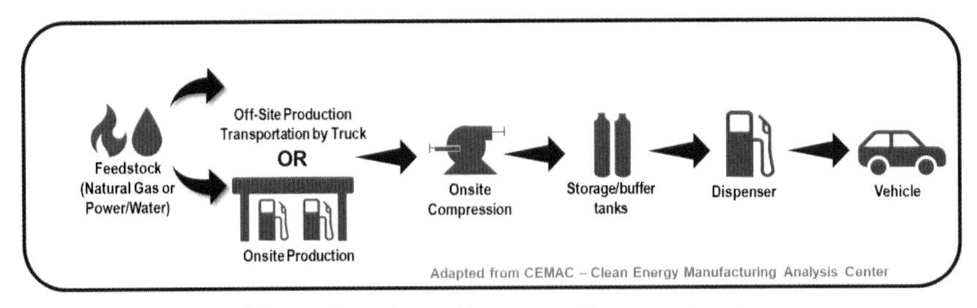

Fig. 12.5 Hydrogen filling value chain. *(Courtesy: RTI International.)*

Ninety five percent of the hydrogen in 2018 was made from natural gas, using a method known as steam methane reforming (SMR). The first portion of the process produces a mixture known as synthesis gas or syngas. This is the building block of most petrochemicals.

$$CH_4 + H_2O \rightleftharpoons CO + 3\,H_2. \tag{12.1}$$

For hydrogen as the end product, an additional reaction is conducted, known as the shift reaction:

$$CO + H_2O \rightleftharpoons CO_2 + H_2. \tag{12.2}$$

The products are hydrogen and carbon dioxide. Note that all the carbon in the methane is converted to CO_2. So, much as in the case of electricity production using fossil fuels, hydrogen production also emits CO_2, albeit not at the tailpipe of the vehicle.

The hydrogen is transported to the filling station. In many cases, a hub-and-spoke model is used, where the hydrogen is first sent to a central facility in liquid form, then distributed to the stations as a gas. Liquid hydrogen transports are not permitted in some inner-city areas. The transport vehicles are known as tube trailers, and the tubes may in some instances be removed and used for buffer storage.

At the filling station, the hydrogen is compressed and stored for dispensing. The pressure will depend upon the pressure of storage in the vehicle but is either 5000 psi or 10,000 psi. These steps of compression and delivery have high capital cost when compared with gasoline pumps.

Also shown in Fig. 12.5 is on-site production of hydrogen. Two methods are feasible. One is electrolysis of water, using line electricity. This can be very expensive and is generally used only in laboratories needing small quantities. A filling station is expected to require 250–300 kg of hydrogen per day. The other technique is small-scale methane reforming, using partial oxidation rather than steam. In such a reducing atmosphere, syngas is produced. It is then shift reacted in a next step to produce hydrogen and CO_2. Ordinarily, such processes require economies of scale to be cost-effective. A new process developed by RTI International using a modified diesel engine as a chemical reactor aims to achieve economical hydrogen production in the required range using natural gas of convenience. This Microreformer™ process was tested at full scale[9] but is not yet in commercial service. Fig. 12.6 compares the delivered cost of hydrogen produced offsite and onsite by appropriate methods. Note the claimed economic advantage of the Microreformer™.

	Off-Site Production & Delivery		On-Site Production		
	Gas	Liquid	On-Site SMR	Electrolysis	Microreformer™
Production + Delivery Cost of 10 bar 99.97% H_2	$6/kg	$8.5/kg	>$7/kg (Best case)	$8.5/kg	Below $4/kg
Onsite Compression, Storage, and Dispensing Cost	$3-4/kg				
Total Cost of H_2 to Vehicle	$9-10/kg	$12/kg	>$11/kg	$12/kg	Below $8/kg
Current Selling Prices of H_2 to Vehicle	$12-16/kg				

Fig. 12.6 Delivered cost of hydrogen produced offsite and onsite. *(Courtesy: RTI International.)*

Offsite production is almost all based on SMR technology. The only variant is in the means of delivery to the user, as a gas or liquid. The cost of SMR is under USD 2.0 per kilogram. The rest of the cost shown is from the delivery, which underlines the need for onsite production. Small-scale SMR plants are being researched. They will probably be housed in hubs, rather than at each station. Ongoing research in electrolytic hydrogen is addressing improved efficiency and cost reduction using a substitute for the costly platinum electrodes.

Hydrogen is sold by weight, not volume. Each car carries on the average 5 kg, and a kilogram contains 40 kWh. The rule of thumb number for miles driven is 70 per kilogram. Driving conditions will affect this. The Toyota Mirai is rated for a range of 315 miles. At 70 miles per kilogram, the vehicle clocks in at 59 MPGe. Had we used the Mirai figure for range, rather than the theoretical 70 miles per kilogram, it would be 53 MPGe. Compare this with the Leaf and Bolt, at 108 and 119, respectively. By this metric, the hydrogen cars are closer to PHEVs and worse than all-electrics by a factor of two. Where, then, lies the allure of hydrogen vehicles?

Comparison of electric vehicles with fuel cell vehicles

The comparison will be made separately for two categories of stakeholder.

Consumers

By all accounts, consumers love the hydrogen version of electric drive. Two key factors seem responsible for this choice: refueling time and range. Filling the tanks takes about 5 min, about what they are used to with gasoline.

The range is 300 miles plus that is well within the comfort zone of the range anxious. No training wheels needed. No changes in behavior, except for one little matter: FCVs are restricted to California for the foreseeable future. A drive to Vegas from San Francisco would not be possible.

For true acceptance, the hydrogen cost will need to descend from the current USD 12–16 per kilogram. At USD 12, this equates to about USD 6 per gallon of gasoline, depending on the vehicle compared. Even in California, that is tough. In much of the rest of the country, gasoline is well under USD 3 and likely to remain there for the long term, given steady oil prices in the USD 50–70 range for the 4 years before the Covid-19 period. Admittedly, a direct comparison with gasoline is not valid because of the other attributes of FCVs, but any lofty goals of elimination of conventional vehicles may require close to cost parity.

However, the real comparison in many ways is with EVs. If fast charging becomes ubiquitous—and this is not unlikely—the convenience factor is addressed. Battery swapping is expected to take no more than 5 min. With all its other advantages, battery swapping's time may have come.

On fuel price, EVs win hands down, especially for now. The FCV consumer is currently being insulated from the high price of the fuel cells through subsidies. Hydrogen is also being subsidized by the car manufacturers, who offer 3 years of free fuel with each lease. That dog will not hunt for long. But recent advances, including the one noted earlier in this chapter, do point to lower fuel prices. The fuel cells are where batteries were 10 years ago. They need similar performance on cost reduction. Toyota appears confident.

Commercial consumers like the fact that the FCV concept applies to virtually all modes of transport, including buses and trucks. Battery-based EVs are hard pressed in these applications.

Suppliers

Automobile companies as a group probably do not have a preference. The bulk of the car is the chassis, electric motors, and control systems. These are nearly the same regardless of the source of the electrons. Because batteries and fuel cells are very different technologies, some auto companies could well be persuaded in one direction or the other if they had proprietary access to technology. The brand-new, from-scratch, company Tesla went batteries and appears unmoved by hydrogen arguments. Founder Elon Musk famously referred to fuel cells as "fool cells." When one has a head start in a technology with economies of scale, one's consideration of doing something that opens up the field is appropriately cautious. However, Musk did quietly make his Model S batteries easily replaceable, even though he did not buy into the Better Place model. This is known as buying optionality. He did not do it with hydrogen.

The other major supplier is of the commodity, electricity, or hydrogen. Electricity is straightforward. Suppliers are extant in each community. Hydrogen is more interesting. A large served market already exists for hydrogen consumed at distributed locations, each using less than 300 kg per day. Note that the FCV filling station requirement is 200–300 kg per day. The size of this already-served so-called merchant hydrogen market is *32 million kg per day*. This is an established industry that has the economics of production and delivery fine-tuned. The leaders are large international players—Linde, Air Liquide, and Air Products, to name the most prominent. They are certainly not ambivalent regarding batteries versus hydrogen.

Hydrogen will probably not make it. Batteries are cost-effective now on capital and operating expense. Charging can be done anywhere, and even fast chargers with wide-bandgap device capability will be priced at under USD 25,000, whereas hydrogen filling units will be closer to USD 1 million. A battery swap model will clinch the deal on charging time and allow for a more sustainable charging infrastructure, using renewables in many cases. Hydrogen fuel cost will come down. But a fuel cell cost decrease may take years. By that time, the installed base of battery EVs may be too large to be supplanted. Also, developing nations such as India do not have an abundance of cheap natural gas (for hydrogen production), whereas utility-scale solar is being commissioned at under 3 US cents per kiloWatt hour. The first vehicles to be electrified in India will be two- and three-wheelers. Battery swapping for them is a simple operation, and so the model may take hold there first. The wild card in predictions is breakthrough technology. Tails wagging dogs.

With little doubt, though, the automotive future is electric.

References

1. Karr J-B; 2020. https://en.wikiquote.org/wiki/Alphonse_Karr. Retrieved April 7.
2. Clark JA, Halbouty MT. *Spindletop*. New York: Random House; 1952.
3. Castelvecchi V, Stove E. Chemistry Nobel honours world-changing batteries. *Nature*. 2019;574:308.
4. Holland; 2018. https://cleantechnica.com/2018/06/09/100-kwh-tesla-battery-cells-this-year-100-kwh-tesla-battery-packs-in-2020/.

5. Heffner RR, Kurani KS, Turrentine TS. Symbolism in California's early market for hybrid electric vehicles. *Transp Res D.* 2007;12(6):396–413.
6. Chafkin; 2014. https://www.fastcompany.com/3028159/a-broken-place-better-place.
7. Iyer VM, Gulur S, Gohil G, Bhattacharya S. Extreme fast charging station architecture for electric vehicles with partial power processing. In: *IEEE Applied Power Electronics Conference and Exposition (APEC), San Antonio, TX, 2018*; 2018:659–665.
8. EnergySage; 2019. https://www.energysage.com/electric-vehicles/buyers-guide/mpg-electric-vehicles/.
9. Rao V, Knight R. *Sustainable Shale Oil and Gas: Analytical, Geochemical and Biochemical Methods.* Boston: Elsevier; 2017:37–42 [ISBN: 9780128103890].

CHAPTER THIRTEEN

Clean coal and dirty solar panels

Particulate matter casts a shadow on solar energy

Solar and wind energy are the most scalable renewable promises for a future with minimal fossil fuel. Of these, solar is the favorite of many low- and middle-income countries, in part because they have very high solar intensities. India has doubled down on solar. The 2022 target of 20 GW was achieved 4 years early. By end of 2019, India had installed capacity of 33.7 GW, and the 2022 target of 20 GW was revised upward to 100 GW. This was enabled by a drop in cost of solar panels, especially from China. The US administration's position on climate change and tariffs on China goods are believed to have been factors. By 2017 Indian solar generation cost was below that of coal-fired generation. Winning bids for utility scale plants were accepted at prices under US 4 cents per kWh.

Particulate matter deposits reduce efficiency of solar panels

This solar juggernaut is facing headwinds from a relatively new realization that airborne particulate matter (PM) is decreasing solar panel efficiency. The role of dust, comprising oxides of silicon and aluminum, has been recognized, and panel washing is a known requirement. Recently, investigators have observed the impact of fossil fuel-derived particulates.[1] The contribution can come from both ambient particle loading and deposition of particles on the panels. In this chapter, we will focus primarily on the latter, it being the dominant contributor.

Transmittance of solar energy to the panel per unit mass of particle is computed by the following equation[1]:

$$\frac{\Delta T}{\mathrm{PM_F}} = -\frac{1}{\mathrm{PM_F}} \sum_{i=1}^{n} (E_{\mathrm{abs},i} + \beta_i E_{\mathrm{scat},i})\mathrm{PM}_{\mathrm{F},i} \tag{13.1}$$

Particulates Matter
https://doi.org/10.1016/B978-0-12-816904-9.00002-7

$\Delta T/PM_F$ is the transmittance per mass loading. There are four components: dust, organic carbon, black carbon, and a grab-bag of other light-scattering ions such as sulfates and nitrates. PM_F is the total mass loading for the period, and $PM_{F,i}$ is the mass loading of component i. The two E's are absorption and scattering efficiencies in m^2/g, and β is the PM upscatter coefficient. The equation importantly assumes a linear relationship between PM mass loading and change in transmittance ΔT, which the investigators believe to be reasonable for the moderate PM loadings in their study.

The study was conducted from January through March on solar panels in Gandhinagar, near the city of Ahmedabad in India. The values for each parameter in the equation are not reproduced here but can be found in the referenced publication. This part of India is known to have substantial ambient PM from anthropogenic sources. Whereas dust is still significant in total mass, those particles tend to be in the coarse range. Fine particles ($PM_{2.5}$, i.e., <2.5 μm in diameter) come from a variety of combustion sources, including coal-fired electricity generation, solid biofuels for heating and cooking, transportation, and roadside trash burning.[2, 3] The ambient levels of PM in these areas often are significantly greater than World Health Organization guidelines, by as much as an order of magnitude.[4]

The left panel of Fig. 13.1 shows the PM components deposited on solar panels in the winter months of February and March. The right panel shows

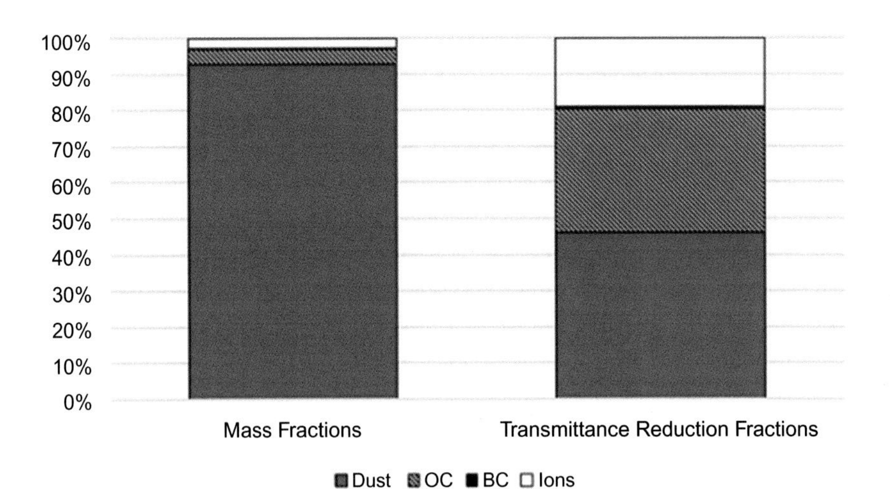

Fig. 13.1 PM components deposited on solar panels and the computed transmittance reduction proportions for each of dust, organic carbon, black carbon, and assorted ionic species.[1] *(Courtesy Bergin MH, Ghoroi C, Dixit D, Schauer JJ, Shindell DT. Large reductions in solar energy production due to dust and particulate air pollution.* Environ Sci Technol Lett. *2017;4:339–344.)*

the relative contribution to transmission reduction by each of the four categories of PM. Even though dust dominates the mass fraction (92%), it contributes only about 40% of the transmittance reduction. More striking is that organic carbon (4%), ionic species (4%), and black carbon (0.01%), together totaling 8%, are responsible for nearly 50% of the transmittance reduction. The graph represents calculations using the equation above, so the results are subject to the values of the parameters. However, the researchers estimated transmittance reduction experimentally from two episodes of solar panel effectiveness measurements several days before and after cleaning. They found the computed value of $-14\%\ g^{-1}\ m^{-2}$ to be within the range of the two measurements of -17 and $-12\%\ g^{-1}\ m^{-2}$.

This finding on transmittance reduction underlines the significance of the disproportionate impact of anthropogenic PM when compared with dust from natural sources. (Wildfires are both anthropogenic and natural, so assigning mass fraction depends on the source of ignition.) For measurements made in Colorado, where the contribution of PM was expected to be primarily dust, the transmittance reduction was $-4\%\ g^{-1}\ m^{-2}$.[5] But this too is substantially less than the $-7\%\ g^{-1}\ m^{-2}$ from dust alone in the current study (deduced from Fig. 13.1 right panel). While differences in dust character could be a factor in the discrepancy, another could be synergistic effects of dust with the other PM.

The direct measurements of solar panel efficiency are striking in the magnitude of the effect of soiling (Fig. 13.2). The electricity generated

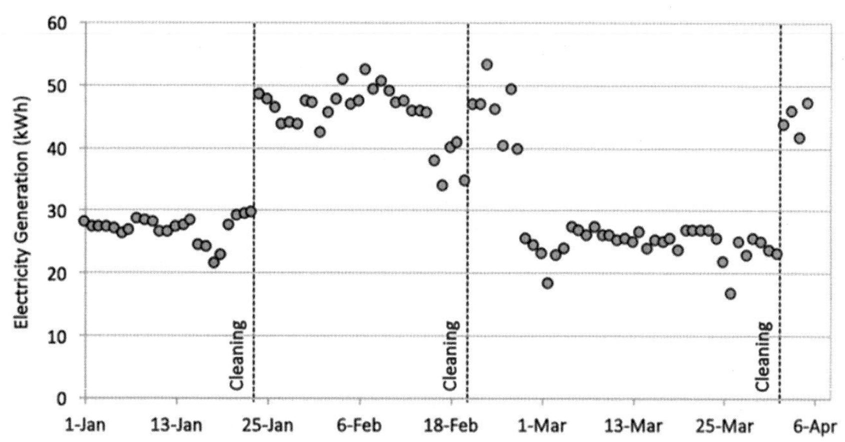

Fig. 13.2 Solar panel efficiency reduction due to PM deposits as related to cleaning events.[1] *(Courtesy Bergin MH, Ghoroi C, Dixit D, Schauer JJ, Shindell DT. Large reductions in solar energy production due to dust and particulate air pollution.* Environ Sci Technol Lett. *2017;4:339–344.)*

jumps by about 50% following a cleaning event. The redeposition and related drop in efficiency appears to occur over a period of 7 or more days. If these results are characteristic of the losses to be expected, a compelling need exists to understand the relative effects of the three nondust species, with a view to controlling their emissions through policy. But even dust, whereas lumped into one category in these computations, is variable in size and chemistry.

Dust is comprised dominantly of sand that is primarily silica (SiO_2) and clay that is a hydrated mixed oxide of Al_2O_3 and SiO_2. A common clay used for pottery is kaolinite, with the formula $2SiO_2{\cdot}Al_2O_3{\cdot}2H_2O$. Some clays have the alkali metals Na or K appended to the molecule, with a very useful such clay being bentonite. Fe is also a component sometimes. The proportion of sand and clay influences the median particle size as well, with sand in general being larger. In the study cited above, the particle size distribution was bimodal, with one mode at 15 μm and the other at 1–2 μm. The latter is clearly associated with the clay component of dust and the combustion-related PM, and most likely is more adherent to the panel than the larger particles.

The disproportionate impact of combustion-related PM on solar panel efficiency points to policy solutions to ameliorate that source, including accommodation to societal norms. These measures notwithstanding, one would need to estimate soiling in real time to clean the surfaces cost-effectively. One investigator demonstrates the efficacy of a low-cost digital microscope to perform this task.[6]

Measuring soiling rates of solar panels

The rate of soiling of solar panels is an important parameter for designing cost-effective cleaning programs. The worst-case scenarios for soiling and cleaning schedules can reduce electricity generation by as much as 35%.[1] But even modest reductions add up to serious income loss. Many panels are highly distributed, such as individual ones on artesian pumps owned by farmers with little sophistication. They need programs supplying guidance on cleaning methods and schedules. A good beginning for such a program is an investigation in western India, which also evaluated a low-cost soiling estimation method.[6] Soiling loss is defined by the following equation:

$$SL = (1 - E_{soiled}/E_{reference}) \times 100\%, \tag{13.2}$$

where SL is the soiling loss for a period and the E's are respectively the energy production from the soiled panel and the cleaned panel over the same period. The low-cost digital microscope was placed in a weatherproof box with a low-iron glass slide providing a window to the microscope. The microscope was pointed at the same angle as the solar panels, 23 degrees due south, 23 degrees being the latitude of the location in Gujarat, India. It shone a light up toward the glass slide and took images every half an hour, although only the night images were usable for analysis due to the glare of the daytime sun.

Fig. 13.3 shows results. It has six panels, each of 3 weeks duration. Panel A has rain events shown in light blue. Shortly after the rain, the PM loading picked up at about the same slope as in the other panels, with a steeper slope early. The uniform increase in loading is logical and very similar for all panels. The observed rate of soiling is 0.37% per day.

Fig. 13.4 demonstrates the linearity of soiling with loading of particulates as estimated with the low-cost digital microscope, with a high correlation, $R^2 = 0.86$. Linearity was taken as an assumption in the earlier referenced Bergin study. This correlation is a strong indicator that the low-cost device is predictive.

The effect of rain on PM deposits is believed to be beneficial. This study appears to show essentially the opposite. The soiling rate during rainy days was more than double that on dry days. The presumptive explanation is that

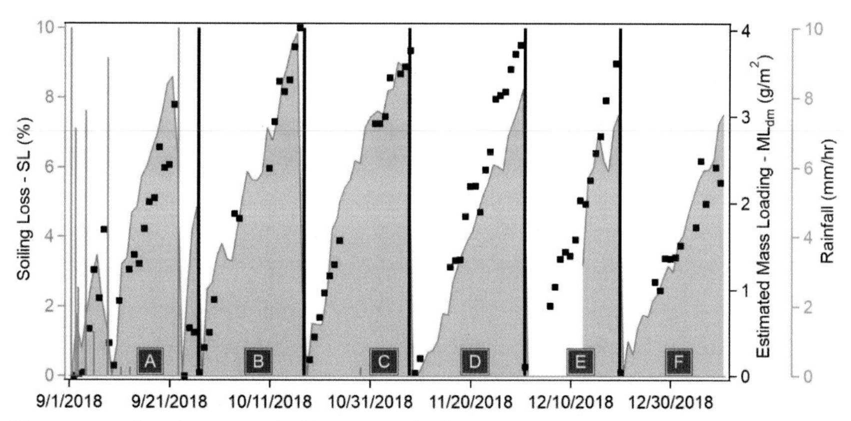

Fig. 13.3 Soiling loss recorded by a Campbell Scientific soiling station *(filled tan spaces)* when compared with collocated mass-loading estimations using a digital microscope *(black squares)*. No soiling station measurements were made from 12/1 to 12/12 (panel E). The *dark vertical lines* represent cleaning events. Rain events are in *light blue vertical lines.*[6] *(Courtesy Valerino M, Bergin M, Ghoroi C, Ratnaparkhi A, Smestad GP. Low-cost solar PV soiling sensor validation and size resolved soiling impacts: a comprehensive field study in Western India.* Sol Energy. *2020;204:307–315.)*

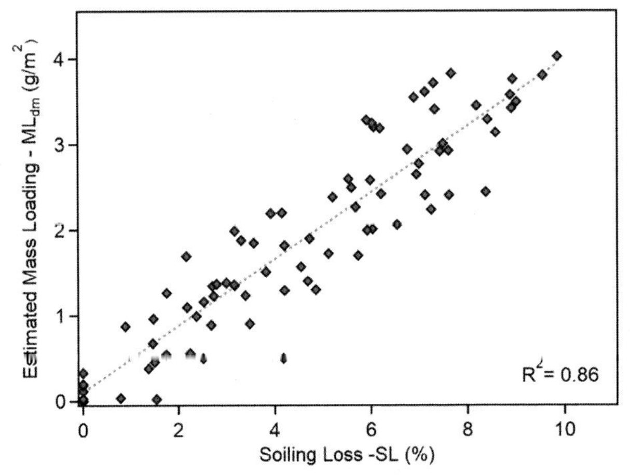

Fig. 13.4 Soiling loss plotted as a function of mass loading estimated using the low-cost digital microscope.[6] *(Courtesy Valerino M, Bergin M, Ghoroi C, Ratnaparkhi A, Smestad GP. Low-cost solar PV soiling sensor validation and size resolved soiling impacts: a comprehensive field study in Western India.* Sol Energy. *2020;204:307–315.)*

rain washes away the larger (>10 μm) particles that tend to be silica and relatively ineffective in diminishing solar panel efficiency. Light rain and high relative humidity appear to increase soiling, most likely through improving adhesion to the panels.[7] Scanning electron microscopy detects the precipitation of adherent crystals. The morphology (not shown here) is indicative of clay minerals. This effect of high relative humidity was observed in multiple locations.[8] The organic carbon (BrC) can be expected to adhere under humid conditions. This mechanism is supported by the observation in panel A of Fig. 13.3, where the postrain rise in soiling loss was noticeably steep.

Adherence is likely a function of both size and chemistry. The evidence points to the larger dust particles, mostly silica, washing off with rain. The particles under 5 μm appear not to be dislodged by dry cleaning that is a factor in water-scarce areas. The Valerino study concludes that much of the soiling impact is by particles in the range of 0.5–1.5 μm (data not shown here). This is consistent with the photovoltaic operating wavelength range of roughly 0.4–0.8 μm; one would expect particles of size comparable to the wavelength to be the most impactful.

A recent study of soiling at seven locations worldwide addressed this issue.[8] They estimated that the greatest transmittance losses were at the shorter wavelengths, as shown in Fig. 13.5 for Chennai, India (the results were surprisingly consistent in all locations). These are wavelengths, where

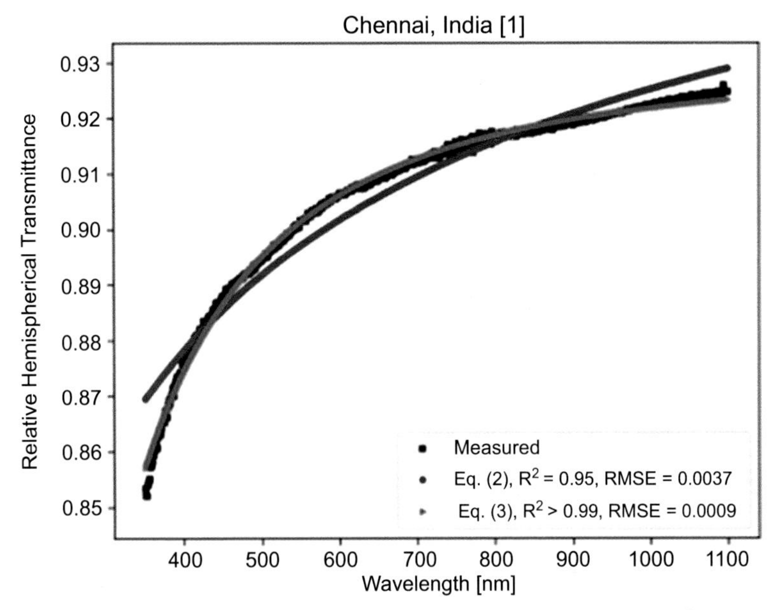

Fig. 13.5 Transmittance in soiled photovoltaic glass panels as a function of wavelength in Chennai, India.[8]

the spectral response for Si solar cells is comparatively low. The likely impact of this observation is that particles approaching the ultrafine range may not be deleterious to solar panel efficiency. This is in stark contrast to the conclusion in much of this book that ultrafine particles are the most worrisome when it comes to health, especially the impact on mortality.

It appears that $PM_{2.5}$ regulations are particularly apropos for the impact on solar energy. $PM_{10-2.5}$ is less relevant, and a move to particle count measurement and associated policy for health applications may not be relevant at all.

Is clean coal indeed an oxymoron?

The press is replete with the opinion, often stated as fact, that clean coal is an oxymoron.[9] No standard definition exists for the term clean coal. Consequently, the assertion could well be made if a very restrictive definition is strictly interpreted. We would support the view "Affordable Clean Coal is an Oxymoron" for many countries. This is because the objectives of clean coal can be met, but at a cost exceeding that of local alternatives. That argument applies to many situations of substitutions or alternatives.

So, the use of the term oxymoron is more for effect than for literal precision. In other words, it is politics. Nevertheless, the issue will be used herein to examine the situation with coal-derived electricity.

Clean coal definition

Into the void created by the absence of a consensus definition we shall plunge. Cleanliness can be ascribed to the material itself, the means of recovery from the earth, the effect on humankind by its combustion process, and the impact of disposal of the products of the process.

Coal comes in four grades, with each having associated qualities affecting the material itself and the means of recovery. All coal was created by the deposition of vegetative matter millions of years ago and its burial under rock overburden. The longer ago the event, the deeper the burial (by and large, unless there were earth movements). Heat and pressure convert the organic matter into coal. The most mature conversions occur at the oldest deposits that are often the deepest. At early ages, the rock (soil) will intercalate in the coal and, when combusted, produce what we know as ash because the minerals remain largely inert in the combustion process (this property is also why the pure versions of these minerals are used to line furnaces to protect the steel bodies). The oldest and deepest deposits will have the least of these ash constituents and the highest carbon content.

Anthracite is the purest form of coal. It is almost glassy in appearance and has between 86% and 97% carbon, almost no water, and very little ash-forming constituents. The other pollutants such as S and Hg are very low. But this is also the least abundant variety, being only 1% of US reserves. Bituminous is the next grade, with carbon content 45%–86%. It is the workhorse coal if available and can have S and Hg. Subbituminous is the next grade down, dipping to 35% carbon. And finally there is lignite that is a very immature form with carbon down to 25%. This last grade can have as much as 35% moisture and 35% ash-forming constituents. The heating value can be as little as half of the top-grade bituminous, largely due to the high water content. In some countries, such as India, lignite is the most abundant grade.

Considering the first two ways of defining cleanliness—material and means of recovery—anthracite is the cleanest on appearance and handling and subbituminous the most friable. But the dust creation propensity aside, this metric is not high on the list. The same goes for mining operations. The best grades are deep and require mine shafts and laterals supported by beams. Danger for miners is present, more so in some countries than others.

Miners used to not have protective equipment, and lung disease ran high in the community (so-called black lung). Conditions are better now. But again, this metric is not high on the list of involved players.

On balance, the clean coal definition appears to rest on the direct emissions at the combustion plant and the disposal of the waste. Emissions fall broadly into two classes, those that have the potential to impair the health of the public in the near term and the long-term effects of emitted CO_2. In general, the public appears more driven by the first class; the principal airborne pollutants are mercury, sulfur dioxide, and PM. The second class is dominated by CO_2, and much of the discourse on clean coal centers on this greenhouse gas.

Close-up: The front of the box

A 2011 *New York Times* story[10] has a very interesting take on the environmental movement and changes in tactics therein. These organizations have in the past targeted national or global goals to appeal to the public. The rise of global ambient temperatures caused by greenhouse gases is a case in point.

The general public can be left cold at two levels. One is that global issues do not resonate with a lot of folks, local ones do. The other is the discounting of future privation. This element of discounting is discussed in Chapter 14. This is not unlike discounting future earnings in finance; a discount rate is applied that gives a lower *present value*. Similarly, future suffering is discounted, especially when it is 40 years out, as are most global warming warnings. Rising water levels on a Florida beach 40 years hence (and only a maybe at that, in some minds) have little resonance with the public in Wyoming. One could call it two degrees of separation.

The *Times* story draws a clever analogy. If a consumer is walking down a grocery store aisle and she sees a box with the image of a brownie on the face, she may be attracted. Some might look at the back of the box detailing information indicative of an obese future for the consumer. Even though the future in this case is more in the short term than the global warming one, the choice of looking at the back is personal and will not happen all the time.

Environmental organizations have been credited with focusing simply on the back of the box, as it were. This stuff is bad for you, we want bigger and bolder signs, we want the warnings to be explicit, and so on. Interestingly, smoking hazard warnings have for years been in front of the box, and probably work better.

According to the *Times* story, some of these organizations are getting the message. They are going local and in front of the box. The first is simply a matter of organization, but the second is a bit harder, because the messaging must hit at the value system. Ocean rise 40 years hence will not play. Cancer risk 10 years down the road will play. Respiratory disease exacerbation now for their children certainly will. So, the Sierra Club focused on individual coal-burning power plants and their presumed impact upon the local population. Shutting these older plants down one by one is the strategy. The efforts have had considerable success and operate in 46 states. In the end, what really worked to secure the shutdowns was natural gas getting reliably cheap. The other option, that of a newer and cleaner coal plant, is not economically justifiable if gas remains relatively cheap. Plentiful shale gas will assure that.[11] At this writing, coal burning plants are on a steep decline. In countries such as India, where the marginal cubic foot of natural gas is from LNG imports, the price is not competitive with coal. But as of 2016, solar power is. When pure economics dictates a sustainable choice, the location on the box no longer matters.

We will mention mercury removal only in passing. Removal is difficult because Hg is in vapor form in the flue gases. Current methods involve converting to a solid compound, adsorbing onto a material such as activated carbon for removal, or selective catalytic reduction. Processes have also been proposed to remove Hg in a pretreatment step designed to beneficiate low-grade coal. One such is microwave treatment, where the water, having high susceptibility, will evaporate and blast out of the coal particles, opening fissures. Hg will volatilize and be collected in much the same way as noted above.

For reasons of focus and brevity, we will discuss primarily the handling of CO_2. The only description will be of the most effective technology for handling sulfur and CO_2, integrated gasification and combined cycle (IGCC). That will be followed by the various means for sequestering the CO_2.

Integrated gasification and combined cycle

Conventional coal-fired plants produce emissions that are handled for removal of contaminants. Fly ash is collected in filtering units in structures known as bag houses. Sulfur compounds are removed at ambient temperature by reaction with solvents. CO_2 capture is done by various means, including adsorption onto an inorganic compound, followed by desorption to collect a concentrated volume. Conventional plants, even the most advanced supercritical (SC) versions, have the pollutants in relatively low concentration. As a consequence, the volume of the carrier fluid treated is high, making the process more expensive. The key difference in the IGCC process is that the combustion is indirect. The coal is combusted in an oxygen-starved environment, resulting in the formation of an intermediate, synthesis gas (syngas), a mixture of CO and H_2:

$$3C + H_2O + O_2 \rightarrow 3CO + H_2 \tag{13.3}$$

This mixture may be combusted to produce CO_2 and H_2O and some uncombusted CO. More commonly, in the IGCC process, the CO portion of the syngas is further shift reacted with water to produce CO_2 and hydrogen:

$$CO + H_2O \rightarrow CO_2 + H_2 \tag{13.4}$$

The hydrogen is combusted to produce power, and water is the only fluid output. Waste heat is recovered after the initial power gas turbine step; this is the "combined cycle" feature that is used in many processes, including natural gas combustion for power.

The syngas is treated to remove sulfur compounds that will be mostly H_2S due to the reducing atmosphere in the reactor. The conventional sulfur-removal methods are at either ambient or chilled temperatures. An interesting innovation removes the sulfur compounds at reactor exit temperatures of up to 950°F by adsorption onto a metal oxide and then desorption at higher temperatures to produce elemental sulfur. The ability to conduct the reaction at higher temperatures reduces the overall cost of the treatment.

CO_2 sequestration

"Sequestration" entails capture and storage, even though the plain language meaning of the term is merely secure storage. In the case of IGCC, the CO_2 is in concentrated form and amenable for any purpose. The most common fate is disposal in a location or form that renders it inert from the standpoint of release to the atmosphere. The other target is beneficial use, giving it a price. The simple disposal has a net cost of transportation to the site and then the cost of the operation.

Storage of CO_2 in a subsurface location is believed to be the most scalable solution to disposal. This process involves injecting the fluid into an oil or gas well for the purpose of storage in a reservoir rock that has been depleted of hydrocarbons. Conventional reservoirs that have produced oil or gas can be expected to have the desirable properties of high permeability (the ability of the fluid to move freely in the rock) and high porosity (proportion of the rock with spaces for fluid). If these parameters are adequate, the depleted reservoir is a good host for the CO_2. The simulations predict 100 years of secure storage. Environmental groups have argued that leakage could occur prior to that, especially through the injector wells. This concern can be overcome by proper selection of the reservoir and maintenance and monitoring of the well.

A more valid concern is the possibility of seismic events when the fluid is injected.[12] Juanes et al.[13] challenged the likelihood of this, arguing that a worst-case scenario injection could cause fractures that allow leakage to the surface through existing faults, but it is probably avoidable. Three-dimensional seismic imaging can identify proximal active faults and their length (length is proportional to the amplitude of the seismic event), and the geology can be chosen to be least susceptible to seismicity. The only issue is scalability: the number of locations that are suitable. This debate notwithstanding, compression and injection have a net cost and, absent inducements such as a carbon tax, will be slow to be adopted.

Mineralization

The concept of fixing CO_2 as a mineral was posited for decades but acquired prominence through the work of Klaus Lackner, then at Columbia University.[14] The mineral would be a carbonate of Ca or Mg, the divalent ions being preferred for fixing more moles of CO_2 than do the alkali metals. In principle, this could be accomplished at some depth below the surface or on the surface and then stored as a stable compound. An issue has been the expected slowness of the reaction. One approach is to do it at depth and not be too concerned about the time for conversion. The other is to catalyze the reaction to be faster on the surface.

The most promising host for mineralization in the subsurface is the rock basalt, with a high proportion of alkaline earth metals. It was formed when magma from volcanic eruptions cooled relatively fast, creating a low-density rock. This is contrasted to slow-cooling species such as granite, which have large crystals and tight-grained texture. Basalt is always fine grained. A variant known as scoria has more than half its volume comprising a highly pitted and porous structure. This was formed when hot gases attempted to escape during the cooling of the flowing lava.

Basalt weathers easily. The dissolution products are alkaline earth (Ca and Mg most importantly) and alkali (e.g., Na and K) minerals. These readily react with carbonic acid in solution created by dissolution of CO_2. The reaction products are carbonates, which are very stable. The common naturally occurring carbonates are limestone and dolomite. Whereas limestone is a structural material, the metamorphic rock derived from it, marble, is even more so and a favorite of sculptors. (Alabaster marble is a misnomer, being a fine-grained $CaSO_4$ rock, not $CaCO_3$; softer and easier to carve than marble, lustrous in its own right but not something to which Michelangelo would have stooped.)

The carbonate crystals grow in the vugs and fissures of the basalt. The reaction is faster if the CO_2 is already dissolved in water. The ocean is an important carbon sink, and one mechanism for fixing the carbon to allow more to dissolve is the reaction with basalt on the sea floor.[15] CO_2 dissolved in water was the means for delivery in a well-reported pilot study in Iceland.[16] They injected CO_2 emissions from a geothermal power plant. The gas was dissolved in water and injected to depths between 400 and 800 m, which is unusually shallow, but perhaps not for Iceland, where the basement rock is probably not very deep. Through interesting ^{14}C tracer means, they were able to show that 95% of the injected CO_2 was retained in the rock within 2 years.[16] They attribute at least some of the rapidity to the gas being in solution when delivered.

A study in eastern Washington state appears to confirm the rates of reaction.[17] They injected 977 tonnes of CO_2 into native basaltic formations and allowed a 2-year soak period. Interestingly, the injected CO_2 was not in solution but in a supercritical (SC) state (see the close-up below for definition). This has relevance to IGCC because the gas captured in the process is essentially pure and can be stored as a liquid. It can later be used, where pure CO_2 is required, as for this study.

Cores were then taken and analyzed. The mineral ankerite—$Ca[Fe, Mg, Mn](CO_3)_2$—was found in vesicles throughout the cores (Fig. 13.6). These form as nodules. In other operating conditions with basalt as host, needle-like crystals were observed.

Isotopes of C and O were used as markers and proved that the carbonate was different from that found in the host rock and had the signatures of the injected CO_2 (Table 13.1). Carbon-containing materials in nature have a distribution of two isotopes, ^{13}C and ^{12}C. ^{13}C has an extra neutron and is heavier as a result. Neither is radioactive (as is ^{14}C that is not an element of this study). The lighter element has a greater vibrational frequency, making it more lightly bound and, therefore, more reactive and ubiquitous. Carbon isotope ratios, shown in the first column, are expressed in parts per thousand deviations from a marine carbonate standard with a ratio of 0.0112372.

The ankerite and the injected gas have nearly the same ratios, as opposed to the native carbonates. This convincingly proves that the ankerite was

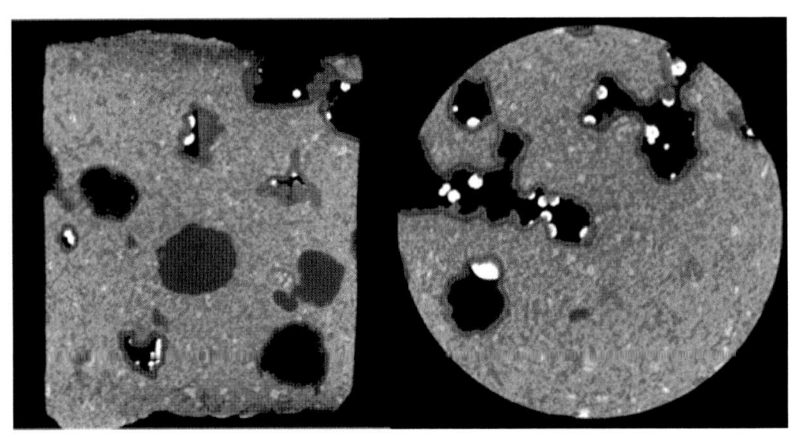

Fig. 13.6 X-ray microtomographic images showing white mineralized CO_2 (ankerite nodules) in the vugs and fissures of basalt in a postinjection sidewall core. Image on the left is a vertical slice and the one on the right is a horizontal slice. *Light gray* is the basalt matrix. *Black* is a void. *(Courtesy McGrail BP, Schaef HT, Spane FA, et al. Wallula basalt pilot demonstration project: post-injection.* Energy Procedia. *2017;114:5783–5790.)*

Table 13.1 Isotopic signatures of pre- and postinjection minerals and fluids.

	Average values (‰)	
Sample ID	$\delta^{13}C$	$\delta^{18}O$
Ankerite nodules	−37.7 ± 2.19	−22.5 ± 2.38
CO_2 source	−36.3 ± 0.09	−27.9 ± 0.51
Formation of water, postinjection	−32.2 ± 0.79	−22.3 ± 1.53
Natural calcite vein	15.8 ± 1.01	−20.9 ± 0.41
Carbonate containing drill cuttings, pre-CO_2 injection	21.3 ± 6.76	−15.1 ± 0.96

sourced from the injected CO_2. Further verification is the isotope ratio of the formation water postinjection, being laden with the signature of the injected gas.

A laboratory study was conducted using core samples from the same Grand Ronde basalt as was used in the injection study noted above. The investigators found mineralization within 20 weeks and concluded that basalt of the Grand Ronde would be a viable site for sequestration, with a loading of 47 kg CO_2 per cubic meter of basalt.[18]

Critics of the Iceland study have raised scalability as an issue, but that is hard to argue, as basalt is one of the more plentiful minerals on the planet. However, much of the coverage is on the ocean floor. One study addresses this point directly. Goldberg et al. identified the necessary attributes of an effective ocean bottom basaltic sink.[19] These include the capacity to hold large volumes in the form of carbonates, with such formations proximal to continental sources, within economical pipeline distances.

Ocean bottoms have many incidences of lava seeps that freeze rapidly into lava pillows that partially disintegrate into high permeability zones.[20] Some of these get overlaid by massive magma sheets that are highly impermeable and that are in turn overlaid by other sediments such as turbiditic sequences (sequential depositions of sand and shale). The net result is a zone of high permeability basalt that can act as a sink for CO_2, overlaid by a trap comprising the magma sheets.

Goldberg et al.[19] identified the Juan De Fuca plate off the west coast of the United States as a site of interest. Wells were drilled and cored. The open-hole wireline logs indicate a more complex structure than generically noted above. Five-to-ten-meter layers of high porosity (10%–15%), as concluded from density and neutron logs, exist between sheets of magma, likely from different volcanic events. They report other drilling studies identifying up to 10-m layers of high porosity aquifers overlaid by impermeable sheets. SC CO_2 injected into these layers could be expected to mix turbulently with the aquifer fluid and be held in place to allow mineralization within the pore spaces and fractures. The stratigraphic trap of the magma sheets is assisted by the likelihood of hydrate formation that, being denser than seawater, will not escape to the surface. This study estimates that the Juan De Fuca plate has enough capacity to sequester 100 years of US production.

The key takeaway from the subsurface sequestering approaches is that injection of SC CO_2 is favored (as opposed to dilute gas), and this directs that coal-burning plants be indirectly fired through the syngas step. The other, rather obvious one, is that there is a net cost associated with capturing, transporting, and storing the CO_2. Studies of fully loaded costs of sequestration on land and in the ocean in sedimentary rock show them to be as low as USD 1 per tonne and as high as over USD 100.[21, 22] Absent regulations, this cost will not easily be borne by the producers. Concern about leakage over 100 years has also been a deterrent. Purely on theoretical grounds, some of which we detailed earlier, this concern ought to be less for basaltic storage because it would apply only to the years in which the CO_2 remained in the gas phase.

Beneficial utilization of CO_2

The most effective form of sequestration is beneficial use. This can take many forms, but all have one aspect in common: someone will pay to use the CO_2. This enables the capture process. The most near-term and proven application is in recovery of oil that will be discussed in some detail because it also entails underground storage. Other areas being investigated include mineralization to produce materials with saleable value and use of CO_2 as a reagent to make chemicals of value.

Enhanced oil recovery

The primary mode of producing oil from reservoirs recovers only up to about 20% of the original oil in place. Secondary methods involve operations such as the use of water, where a "flood" of water pushes residual oil toward receiving wells. This sweep mechanism can recover another 10%–20%. Typically, at the end of secondary operations, about 65% of the original oil remains in the reservoir.

The tertiary oil recovery method of interest to us uses CO_2, preferably sourced from industrial emissions. This involves the injection of SC CO_2 into the formation. As noted earlier in the chapter and in the close-up below, this fluid has the unique attributes of a liquid and a gas. With gas-like character, it can enter the pores in which the residual oil resides. Then the liquid-like character kicks in, and it mixes effectively with the oil phase. Being over twice as soluble in oil as in associated water helps target the oil. It causes oil swelling and reduced oil viscosity even prior to the mixing. The mixing aspect is why the operation is labeled a miscible flood. This nomenclature distinguishes it from the immiscible water flood type of operation.

Fig. 13.7 describes a typical operation. The depicted one has alternating floods of CO_2 and water, but just CO_2 is commonly used. The CO_2 will be compressed to be in SC form. The technique is most effective when the pressure exceeds the minimum miscibility pressure. This is defined as the pressure at any given temperature above which miscibility is achieved. In typical oilfield conditions, it is achieved at depths exceeding 2500 ft. Necessarily, this is just a rough figure, the actual value relying on the pressure gradient in the area, but the point is that the depth is easily exceeded for most producing reservoirs in conventional oil fields. The mixture of oil and CO_2 has a reduced interfacial tension and increased volume when compared with the oil in place. The continuous injection imposes a pressure drive that pushes the mixture toward the producing well.

The produced mixture is processed to release the CO_2 that is compressed and reused. Only a portion of the injected CO_2 mixes; the rest remains in the reservoir. These proportions vary, but eventually 95% of the injected CO_2 remains in the formation and is subjected to the conventional trapping mechanisms for retention. This sequestration feature makes SC CO_2

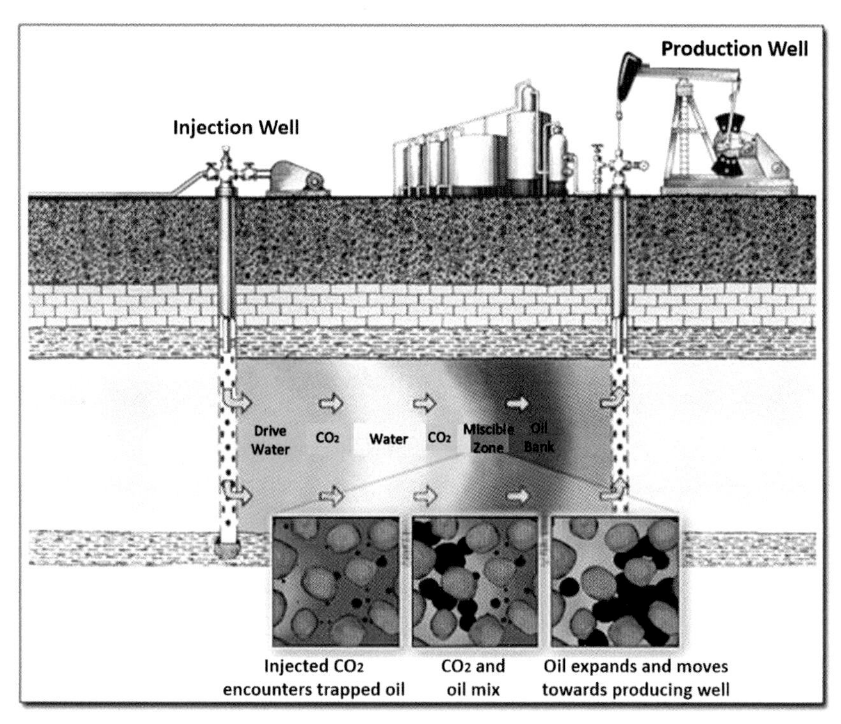

Fig. 13.7 Typical CO_2-miscible flood operation. *(Courtesy United States Department of Energy.* Storing CO2 with Enhanced Oil Recovery. *Washington, DC: Department of Energy, National Energy Technology Laboratory; 2008. Prepared by V. Kuuskraa, R. Ferguson, Advanced Resources International, DOE/NETL-402/1312/02-07-08.)*

injection for enhanced oil recovery (EOR) compelling: a beneficial use paying for the operation also results in storage.

Enhanced oil recovery economics and potential role in CO_2 sequestration

The obvious criticism of the use of CO_2 in EOR is that it abets perpetuation of fossil fuels. Most coins have an obverse side. Here, we will discuss whether a net positive emerges from EOR.

One factor to be considered is that oil is getting increasingly difficult to find and develop. Not counting shale oil, which has its own issues, most of the unexploited reserves in the world are heavy oil. The production of heavy oil carries its own carbon footprint concerns, largely in the production of the steam required to recover it. A conventional reserve already has sustained the environmental impact of operations on the location. One could well argue that producing more from an existing location is more environmentally attractive than pursuing new areas. This is especially in focus in the United States now, where the current administration is considering opening the Alaska National Wildlife Refuge. Some estimates have it that the economically recoverable EOR oil in the United States is over 50 billion barrels at the USD 70 price, when compared with an estimated 15 billion barrels reserve in the Refuge.

The oil produced through EOR has a different carbon footprint than conventional oil. This was computed by Nunez-Lopez and Moskal using a life cycle analysis and assuming that all the CO_2 was derived from an industrial process such as a power plant.[23] They conclude that the oil recovered by EOR represents a net CO_2 emission reduction of 37%. This is another plus for SC CO_2-based EOR over new production.

Another argument in favor is that SC CO_2-based EOR could be training wheels for widespread sequestration in depleted reservoirs. The concept is that the infrastructure built for EOR could be leveraged for later use for conventional sequestration in either saline formations near the one being operated for tertiary recovery and/or that reservoir after it is finally depleted.[24] They refer to this as the "stacked reservoir" concept. But the key is the existence of the pipelines bringing the CO_2 and, almost as important, the existence of a compression facility.

Although the precise figure is reservoir dependent, the CO_2 required for EOR ranges from 300 to 600 kg per barrel of oil recovered.[25] For purposes of calculation of economics, we will take 450 kg, that is 0.225 tonne. At an oil price of USD 70, each barrel produced would require the use of 0.225 tonne CO_2, costing USD 11.25, assuming a CO_2 price of USD 50 per tonne. This cost is likely sustainable even without subsidies, in part because it is what the industry refers to as the marginal barrel of oil. The cost of oil is divided into exploration to locate the reserve, development to drill wells in locations to drain it, and lifting that is the cost of pumping it up. The first two have already been expended in the case of a field destined for EOR. Arguably, the lifting cost is a bit higher because of the compression facility. But the overall margin on USD 70 oil should be able to handle a USD 11 cost for the CO_2.

Then, there is the prospect of subsidies through policies to encourage sequestration. In the United States at this writing, the policy is guided by the 45Q Federal Tax Credit, produced from bipartisan action in 2018. The relevant elements are

- USD 35 credit for CO_2 from anthropogenic sources used for EOR.
- USD 50 credit for CO_2 from anthropogenic sources used for carbon capture and storage outside of EOR.

This credit is available to the party that captures the CO_2, which in this case is the industrial facility. It is available only if the party ensures the execution of the storage in either of the categories (which does include other beneficial uses of CO_2). The capture party may share the credit with the storing party.

In the example above, the oil company would pay USD 50 for the product, and the entity capturing would need to be able to do it for USD 85. For capture costs, these figures are in a sweet spot allowing the business to proceed and using the 12 years (current duration of the subsidy beginning at the capture plant commencement of service) to bring costs down. Those that bring costs down faster could share some of the subsidy with the oil company and improve market share.

The EOR business could be seen as a steppingstone to other sequestration because it enables the cost reduction in the capture. As the subsidy is higher for non-EOR use, the stacked reservoir concept could be even more viable. In some ways, the cost reduction in solar panels driven by the demand created by Germany's policy on solar energy can be viewed as an analog for what EOR may achieve in CO_2 sequestration. For that reason alone, people fundamentally opposed to fossil fuels should hold their powder dry.

Another close-up: Supercritical CO_2, *Oliver Twist*, and coffee

Carbon dioxide exists in three stable states: gas, liquid, and solid. Temperature and pressure determine the state. Fig. 13.8 is the phase diagram.

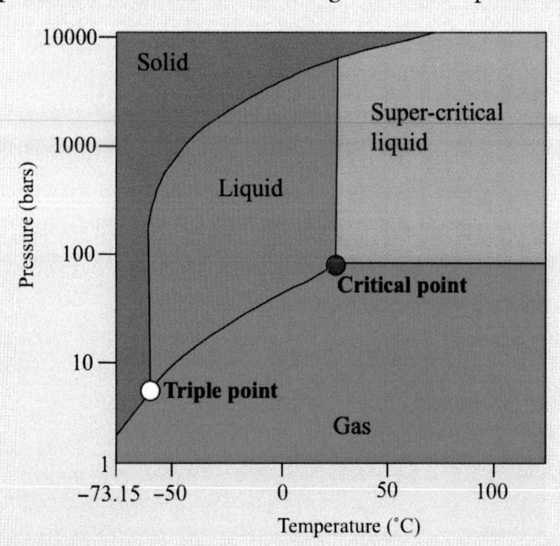

Fig. 13.8 Phase diagram for CO_2 identifying the supercritical zone. *(Modified from Geological Digressions; June 2016. https://www.geological-digressions.com/carbon-capture-and-storage-geological-traps/. Retrieved 17 February 2020.)*

The triple point (white dot in the figure) is the point in temperature and pressure space, where the three phases are in equilibrium. The critical point (red dot) is at a pressure of 73.9 bar and a temperature of 31.3°C. At temperatures and pressures higher than that, the fluid is in both a liquid and gas state. It will expand to fill a volume, as does a gas, but will have the density of a liquid. It will enter spaces like a gas but will behave like a liquid in reacting with constituents to dissolve them. This property is used to extract various chemicals and, in an interesting application, as a dry cleaning agent. It is also the attribute well suited to mineralization of CO_2 in basaltic vugs and fissures. In this and other sequestration applications discussed in this chapter, the fluid will generally be in the lower left quadrant of the yellow zone, close to the critical point. This is also the state in which it is injected in oil wells for tertiary recovery.

Oliver Twist

A London scene in any enactment would not be complete without billowing white fog invading the stage close to the floor and then spreading. This effect is achieved by dropping pieces of solid CO_2 (dry ice) into a vessel containing boiling water. Because CO_2 cannot be in a liquid state at atmospheric pressure at that temperature (see phase diagram), the dry ice sublimates directly to gas that produces the desired visual effect.

Decaf

The principal allure of coffee is the flavor, but the wakefulness induced by it is also prized. This latter is attributed to caffeine. According to legend, an Ethiopian goatherd noticed that goats eating beans from a bush were unable to sleep. Likely apocryphal, the location is about right because coffee is believed to have originated in that region and first concocted into a brew in Arabia, with further legends associated with that event, provenance equally doubtful.

Late night coffee is not for everybody. A minority are unaffected. The rest lose sleep, as the mythical goats did back in Ethiopia. Coffee manufacturers, ever willing to serve, stepped up with processes that took out the caffeine. They named the product decaf.

All treatments are to the green beans, not roasted ones. (In an interesting twist of nomenclature, all have a greenish hue.) In the early going, they used organics such as benzene, a real showstopper, and later ethylene acetate, relying on the fact that a few fruits have small quantities of this stuff and so, of course, a "natural" label was available. More recently, the Swiss Water process is the go-to method. Hot water extracts both the caffeine and the flavorful ingredients. The laden water is filtered through activated charcoal, where the larger molecules of caffeine are trapped. The remainder is used on a fresh batch of green beans (the denuded batch is discarded). Because the once used water is saturated in the flavorful molecules, they will not transfer to the water from the beans and only the caffeine is extracted, leaving behind flavorful beans after removal of the water.

The most recent process uses SC CO_2. The green beans are first steamed and swell to almost twice their size. Then, SC CO_2 is injected at high pressure. It reacts preferentially with the caffeine, leaving behind the flavor-rich ingredients. The caffeine-laden fluid is removed and passed through a charcoal filter to capture the caffeine, and the CO_2 is recovered for reuse. The caffeine is extracted from the filters and sold, for example, to soft drink manufacturers.

And the consumers of the resultant coffee can slip into the clutches of Morpheus.

References

1. Bergin MH, Ghoroi C, Dixit D, Schauer JJ, Shindell DT. Large reductions in solar energy production due to dust and particulate air pollution. *Environ Sci Technol Lett.* 2017;4:339–344.
2. Villalobos AM, Amonov MO, Shafer MM, et al. Source apportionment of carbonaceous fine particulate matter (PM2.5) in two contrasting cities across the Indo-Gangetic Plain. *Atmos Pollut Res.* 2015;6:398–405.
3. Vreeland H, Schauer JJ, Russell AG, et al. Chemical characterization and toxicity of particulate matter emissions from roadside trash combustion in urban India. *Atmos Environ.* 2016;147:22–30.
4. Dey S, Di Girolamo L, van Donkelaar A, Tripathi SN, Gupta T, Mohan M. Variability of outdoor fine particulate (PM2.5) concentration in the Indian subcontinent: a remote sensing approach. *Remote Sens Environ.* 2012;127:153–161.
5. Boyle L, Flinchpaugh H, Hannigan MP. Natural soiling of photovoltaic cover plates and the impact on transmission. *Renew Energy.* 2015;77:166–173.
6. Valerino M, Bergin M, Ghoroi C, Ratnaparkhi A, Smestad GP. Low-cost solar PV soiling sensor validation and size resolved soiling impacts: a comprehensive field study in Western India. *Sol Energy.* 2020;204:307–315.
7. Mekhilef S, Saidur R, Kamalisarvestani M. Effect of dust, humidity and air velocity on efficiency of photovoltaic cells. *Renew Sustain Energy Rev.* 2012;16(5):2920–2925.
8. Smestad GP, Germer TA, Alrashidi H, et al. Modelling photovoltaic soiling losses through optical characterization. *Sci Rep.* 2020;10:58. https://doi.org/10.1038/s41598-019-56868-z.
9. Dorner J; 2017. https://www.huffpost.com/entry/brian-williams-clean-coal_b_144764.
10. Kaufman L. Environmentalists get down to earth. *New York Times.* December 17, 2011.
11. Rao V. Gas positioned to displace coal plants. *American Oil and Gas Reporter.* June 2011.
12. Zoback MD, Gorelick SM. Earthquake triggering and large-scale geologic storage of carbon dioxide. *PNAS.* 2012;109(26):10164–10168. https://doi.org/10.1073/pnas.1202473109.
13. Juanes R, Hager BH, Herzog HJ. No geologic evidence that seismicity causes fault leakage that would render large-scale carbon capture and storage unsuccessful. *PNAS.* 2012;109(52):E3623. https://doi.org/10.1073/pnas.1215026109 [issue of December 26].

14. Lackner KS, Wendt CH, Butt DP, Joyce EL, Sharp DH. Carbon dioxide disposal in carbonate minerals. *Energy*. 1995;20:1153–1170.
15. Alt JC, Teagle DAH. The uptake of carbon during alteration of ocean crust. *Geochim Cosmochim Acta*. 1999;63(10):1527–1535.
16. Matter JM, Stute M, Snaebjornsdottir SO, et al. Rapid carbon mineralization for permanent disposal of anthropogenic carbon dioxide emissions. *Science*. 2016;352:1312–1314.
17. McGrail BP, Schaef HT, Spane FA, et al. Field validation of supercritical CO2 reactivity with basalts. *Environ Sci Technol Lett*. 2017;4(1):6–10. https://doi.org/10.1021/acs.estlett.6b00387.
18. Xiong W, Wells RK, Horner JA, Schaef HT, Skemer PA, Giammar DE. CO2 mineral sequestration in naturally porous basalt. *Environ Sci Technol Lett*. 2018;5(3):142–147.
19. Goldberg DS, Takahashi T, Slagle AL. Carbon dioxide sequestration in deep-sea basalt. *Proc Natl Acad Sci U S A*. 2008;105:9920–9925.
20. Fisher AT. Permeability within basaltic oceanic crust. *Rev Geophys*. 1998;36:143–182.
21. Eccles JK, Pratson L. Economic evaluation of offshore storage potential in the US exclusive economic zone, greenhouse gases. *Sci Technol*. 2013;31:84–95.
22. Eccles JK, Pratson L, Newell RG, Jackson RB. Physical and economic potential of geological CO2 storage in saline aquifers. *Environ Sci Technol*. 1969;43(6):1962–1969.
23. Núñez-López V, Moskal E. Potential of CO2-EOR for near-term decarbonization. *Front Clim*. 2019;1:5. https://doi.org/10.3389/fclim.2019.00005.
24. Nagabhushan D. *Leveraging enhanced oil recovery for large-scale saline storage of CO_2*; 2019. https://www.catf.us/2019/06/leveraging-enhanced-oil-recovery-for-large-scale-saline-storage-of-co2/.
25. McGlade C. *Can CO2-EOR really provide carbon-negative oil?* 2019. https://www.iea.org/commentaries/can-co2-eor-really-provide-carbon-negative-oil. Retrieved 19 February 2020.

PART FIVE

Policy directions

The earlier sections described how PM is created and measured. Also described were health effects and engineered interventions to ameliorate or eliminate the injurious consequences of anthropogenic PM. This section recognizes that interventions are influenced and driven by policy and associated regulations. Regulations are informed, in turn, by epidemiological research. Such research is lacking in high emissions areas of the world.

Household air pollution is not regulated, despite being the largest contributor to PM mortality. Policies offering inducements for safer practices may be more effective than regulations. Much evidence points to the ultrafine portion of $PM_{2.5}$ being disproportionately harmful. Research is required on proving causality. This ought to be followed by regulatory policy.

CHAPTER FOURTEEN

Research and policy directions

This chapter looks to the future in two principal intersecting areas. One is research in mechanisms of diseases created or fostered by particulate matter (PM). Included are studies of the health of populations under different conditions of ambient pollution. The other is the development of policy in assuring public health, largely informed by disease-focused research, especially epidemiological studies connecting mortality to degrees of pollution.

A reason for emphasizing these two topics is that much remains to be done. One could safely make that comment about any disease and its causes. What, then, makes airborne particulates different? PM is the single biggest public health problem in the world.[1] It is responsible for over 6.7 million avoidable deaths per year, breaking down into 3.2 million and 3.5 million for outdoor airborne pollution and household air pollution (HAP), respectively.[2] Most of the latter are in low- and middle-income countries (LMICs). The HAP category is essentially nonexistent in the United States and Europe because the principal contributor is biomass burning for cooking. In some sub-Saharan African countries, over 90% of the population is subjected to HAP.

To complicate matters, almost all the studies connecting ambient air pollution with mortality have been done in the United States and Europe,[3] where ambient PM is much lower than in LMICs that regularly have ambient PM *many times* the mean levels in the United States. Even more striking is the exposure levels for cooks using biomass cookstoves in indoor settings. PM levels up to 950 μg/m^3 were measured in one study in Sri Lanka.[4] Modeling is necessary to extrapolate to these levels. We will be discussing the uncertainties in such extrapolation and the fallacies in the assumptions of linearity. We report on recent gains in that space, while also underlining the need for further research, not just in the modeling, but also in the conduct of epidemiological studies in areas of high exposure.

Policy revision needed

The title of this section might appear at first blush to be a statement of the obvious. All policies in any space ought to be reviewed periodically for

Particulates Matter
https://doi.org/10.1016/B978-0-12-816904-9.00011-8

changes in the underlying conditions informing the policy. Carbon mitigation is viewed as a global issue, and countries attempt to coordinate policy, albeit with limited success. PM should be viewed in the same way, in part because transmigration is not in dispute even at distances traversing the Pacific, as described in Chapter 1.[5, 6] It is often a significant contributor to the ambient count in places that follow strict control practices.[7] For example, in the winter monsoon season in Taiwan, pollution from mainland Asia contributed 30 μg/m^3 to the ambient count.[8]

Disparity in regulations is a compelling reason for all jurisdictions to take a hard look at them, as is evident in Table 14.1. Each jurisdiction also has figures for daily means that are always higher than the annual means. For example, the World Health Organization's (WHO) figure for a 24-h mean is 25 μg/m^3. In some jurisdictions, such as Saudi Arabia, much of the PM is from desert sand and, therefore, not anthropogenic.

Unsurprisingly, the two countries with the highest PM-related mortality, China and India, have the highest thresholds. But Pakistan, weighing in at the number 3 position on mortality, sports a 15-μg/m^3 threshold. One conclusion is that the country is not meeting the threshold in practice. One study estimated the change in worldwide mortality for 2010 if all countries had adopted the European Union or US standard: a 17% reduction if the European standards were used, 46% if the US standards were used.[9]

Table 14.1 Regulatory levels of annual means for $PM_{2.5}$ in selected countries and unions.[9]

Country or union	$PM_{2.5}$ annual mean (μg/m^3)
United States	12
European Union	25
Canada	10
Mexico	15
China (Beijing)	35
Australia	8
India	40
Pakistan	15
Bangladesh	15
Saudi Arabia	15
World Health Organization	10

Courtesy: Giannadaki, Lelieveld J, Pozzer A. Implementing the US air quality standard for $PM_{2.5}$ worldwide can prevent millions of premature deaths per year. *Environ Health*. 2016;15:88. https://doi.org/10.1186/s12940-016-0170-8.

This illustrates the nonlinearity of the effect, with disproportionately greater gains being achieved at the lower ranges.

Most investigators assessing mortality risk use a variant of the formula

$$\Delta\,\text{Mortality} = \text{Pop}\left(1 - e^{\beta \times \Delta X}\right) Y_0, \qquad (14.1)$$

where Pop is the population, Y_0 is the baseline mortality, ΔX is the change in the annual mean $PM_{2.5}$ concentration, and the β coefficient is the concentration-response factor derived from the relative risk determined from many epidemiological studies. The connection to actual deaths is solely through the epidemiological studies. Accordingly, depending on which studies are relied upon, estimates could differ. This will account for the fact that slightly different figures may be cited in different parts of this book.

Zhang et al. estimated mortality as a function of ambient $PM_{2.5}$ experienced over two decades.[10] They and others in recent studies focus on ambient pollution ranges in the United States, which are below 30 μg/m^3.[2, 3, 11] They all present the data with linear fits. If these linear profiles and slopes were to be extended to ambient loadings in some Asian cities (see Table 1.2 in Chapter 1), which are over 100 μg/m^3, the predictions of mortality would be nonsensical in some cases (over 100% mortality). Clearly, therefore, the curves flatten at high ambient numbers. The need for extrapolation is driven by the fact that no epidemiological studies exist connecting long-term exposure to direct measurements of $PM_{2.5}$ with cardiovascular and respiratory disease caused mortality in regions of the world with high ambient $PM_{2.5}$ exceeding 100 μg/m^3.[3] Certainly, the HAP sector is also severely underrepresented in this regard, despite overall mortality worldwide being higher in these environments than in all other walks of life combined. HAP environments also experience very high $PM_{2.5}$ concentrations, well in excess of 100 μg/m^3.

Facing this deficit in epidemiological studies at high PM levels, Pope[12] suggested using investigations of the health effects of smoking, both active and passive, to shed light on the shape of the exposure-response curve at higher ambient $PM_{2.5}$ exposures. This was enabled by the fact that smoking-based exposures covered a wide swath up to very high loadings representative of hundreds of micrograms per cubic meter of ambient $PM_{2.5}$ loading from other sources. Necessary was the assumption that the nature of inhalation did not matter, nor did the variability in the material being combusted. This last we know now not to be accurate, in that the organics produced are different for different combusted matter, and that the organic layers play a prominent role in toxicity of the particles. However, that detail likely does not affect the broad contours of the response curve at high PM levels.

Nonlinearity of mortality risk with increasing $PM_{2.5}$ loading was a critical finding in the investigation of cigarette smoke as a cause for lung cancer and cardiovascular diseases. A very steep rise in mortality risk occurred at relatively low dosing (smoke inhalation), and at higher daily exposures to cigarette smoke, the slope changed from linear to a flatter profile.[12] Also plotted on the curves were mortality estimates from long-term ambient exposure to $PM_{2.5}$. Because all the ambient exposure investigations were at relatively low $PM_{2.5}$ levels, they were in the linear portion of the curve. However, the point got made that they generally fit the model.

Fig. 14.1 shows the results, with the vertical axis plotting the adjusted relative risk of ischemic heart, cardiovascular, and cardiopulmonary mortality against exposure to cigarette smoke, both passive and active. Data from various studies of exposure risk for ambient air pollution are at such low levels when compared with the cigarette smoke exposures, that they had to be shown in an exploded view in the inset. Similar results were obtained for the adjusted relative risk for lung cancer mortality, with the flattening not as pronounced (data not shown).

Pope's original point, the use of smoking data as a window to nonlinearities at high ambient PM concentrations, appears to have been made. The only questionable assumption is that PM from combustion of cigarettes is similar enough in toxicity to be compared with ambient PM from myriad sources. But this remains a viable hypothesis because the linear extrapolation to very high levels of ambient PM gives impossible results. Underlined is the need for epidemiological studies at the high levels.

Assuming that the type of risk-exposure profile shown in Fig. 14.1 applies to ambient $PM_{2.5}$, the implications for policy appear fraught. A flat profile at high exposures would appear to suggest that reductions at the highest levels will produce small reductions in mortality. This further suggests that the cost of the reductions may exceed the economic benefit and that substantial movement in ambient pollution may be needed to make a material impact on disease reduction. One publication addresses this potentially troubling conclusion in some detail, using examples of pollution reduction in India and China.[2] The key point is that the marginal benefits of policies that reduce ambient air pollution, even at the high levels with near-flat profiles, depend on three parameters. These are the size of the affected population, the baseline mortality rate, and the value ascribed to each death reduction. The conclusion is that, the flat profile notwithstanding, substantial economic benefits are to be derived from incremental reductions even in areas with high ambient pollution. One example is in the

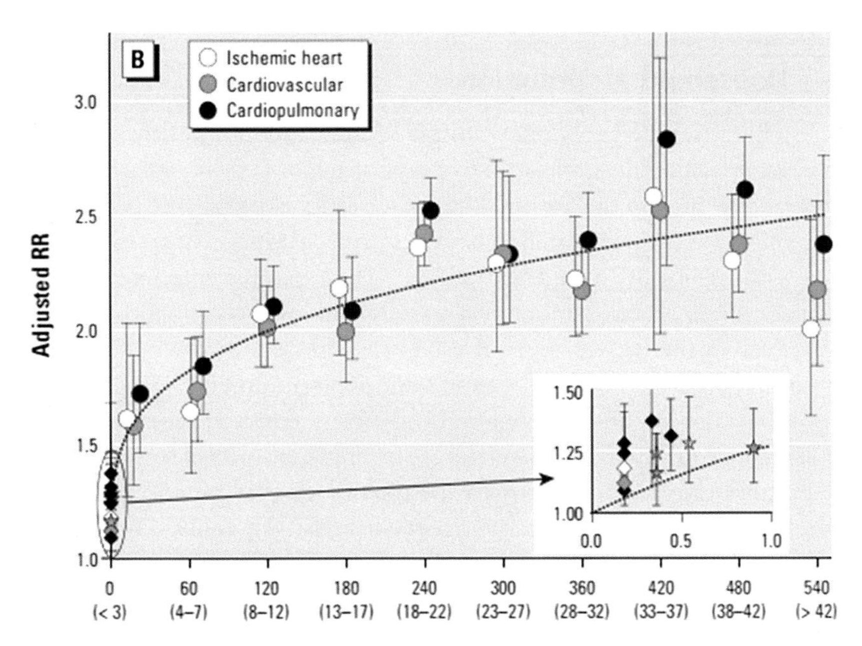

Fig. 14.1 Adjusted relative risk of ischemic heart, cardiovascular, and cardiopulmonary mortality versus daily cigarette smoke inhalation, both active and passive. The numbers in parentheses are cigarettes per day. Also plotted are data from studies of ambient air pollution. The PM loading in cigarette smoke inhalation is greater when compared with that from ambient air pollution; therefore, the ambient PM data are clustered in the near-vertical portion of the curves near the vertical axis. The detail inset reproduces those data for clarity; the *dotted line* is the nonlinear power function going through the origin.[12] *(Courtesy Pope III CA, Burnett RT, Turner MC, et al. Lung cancer and cardiovascular disease mortality associated with particulate matter exposure from ambient air pollution and cigarette smoke: shape of the exposure–response relationships.* Environ Health Perspect. *2011;119:1616–1621. https://doi.org/10.1289/ehp.1103639 (web archive link).)*

desulfurization of flue gases from coal fired plants in India. In a study of 72 plants of varying sizes, the authors conclude that the economic benefit exceeds the cost, especially in densely populated areas.[13]

Another consideration is that even in relatively less-polluted nations, pockets of high ambient pollution coexist with proximal low pollution areas, often rural. Transmigration (see Chapter 1) is a reason to reduce levels in the most polluted pockets.

Household air pollution

This area is certainly one with the greatest need for policies that regulate, or at least guide, emissions. Yet regulations are absent, even in countries with the highest disease outcomes. One rationale is that the polluters, if you will, are individual families, largely in rural settings. Assuring compliance with regulation would be difficult in part because interference with a way of life in a poor community is inadvisable without the provision of cost-effective alternatives. Even if such were devised, they would need to be strongly informed by local customs and societal mores.

Ambient air pollution is predominantly from industrial sources. Eighty-six percent of all airborne particulates are from combustion of some sort, and the top three are coal for electricity, petroleum for transportation, and biomass. The first two are industrial and regulated relatively easily. They are also well suited for engineered interventions because of the concentration of activity. In contrast, HAP is difficult to influence. Much effort has been expended for decades on improving the efficiency of stoves (see Chapter 8). This certainly lowered emissions per meal cooked, but the levels remained high from the standpoint of morbidity and mortality. The basic problem was that the reduction in disease outcome with reduced exposure was nonlinear.

A clue to the reason appeared in the modeling of data from all forms of combustion, including tobacco smoke, both active and passive. This last has some of the highest exposures of any source other than HAP that also runs high depending on the type of stove and environment. Burnett and others estimated the impact of $PM_{2.5}$ on acute lower respiratory infection (ALRI) in children under 5 years of age; the relative risks for ALRI were estimated as a proxy for "mortality and lost-years of healthy life" in these children (Fig. 14.2). They concluded that the entire impact was from HAP because the children would not be active smokers. In many LMICs, the incidence of ALRI in young children has been reported in connection with cooking with biomass. The cooks, almost always women, will have the children next to them during the cooking.

The modeled curve underpredicts the HAP data, but the important takeaway is that the curve tends to flatten at high exposures. Because ALRI can serve as a proxy for mortality in children under 5 years of age, this figure implies that with higher levels of PM, we can expect a flattening mortality curve for those who are exposed mainly to HAP. Note the high values for

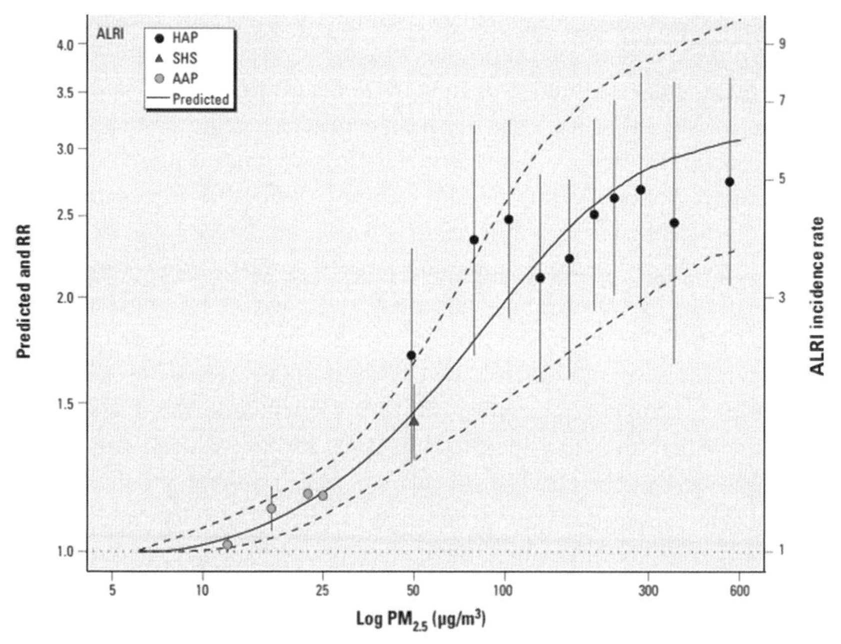

Fig. 14.2 Predicted model for relative risk (RR), *solid line*, with *dashed lines* showing the 95% CI. The type-specific RRs are the *solid circles*; the *bars* show the 95% CI, with the HAP data in *red* and ambient air pollution in *green*.[3] *(Courtesy Burnett RT, Pope III CA, Ezzati M, et al. An integrated risk function for estimating the global burden of disease attributable to ambient fine particulate matter exposure.* Environ Health Perspect. *2014;122:397–403. https://doi.org/10.1289/ehp.1307049 (web archive link).)*

HAP exposure, consistent with results reported in the Sri Lanka study.[4] Note also that it can be as low as 50 μg/m^3 with some form of intervention, likely improved stoves with chimneys. The curve appears to be asymptotic at the low numbers but that is unlikely in practice. Considerable evidence exists pointing to linear gains down to zero.

The flattening at high exposures indicates the need for radical measures to attack the problem. The most prevalent of these has been the attempted substitution of liquefied petroleum gas (LPG) or electricity for woody biomass. Because most of the affected population is rural, electricity is usually expensive and in short supply. In most of these localities, electricity is not sufficient even for the basic needs of lighting, fans, and cell phone charging. Consequently, LPG is the substitute of choice in India. Federal subsidy programs exist for the fuel, and there is also some support for the cost of the stove. But the system is plagued by uncertainty of fuel supply. Most importantly, the cost is prohibitive. Even when LPG is installed, many households continue to use wood from some to a great extent, a practice known as stacking.[14] An LPG cylinder in India holds 14.2 kg of gas and has a tare weight of about 15 kg. Villagers, literate or not, are reluctant to demand weighing of the cylinders upon receipt, even though the scales are required to be present in the distribution depots. The result is that unscrupulous merchants short each cylinder, often by just 60 g, likely not noticeable in a total 30 kg object,[15] yet profitable in volume.

In most villages, the householder must travel to the distributor to pick up the full cylinder because the distributor will make deliveries to a given village only on a schedule, and no household can afford to have a spare tank. In a survey conducted in the Indian state of Bihar, the percent usage of LPG in households was inversely proportional to the distance to the distributor.[16] Distances ranged from 4 to 40 km. The survey also found that many households did not get the subsidy paid to them because literacy deficits made them unable to negotiate the procedural hurdles. Considerations such as these are the type of intangibles that are community specific, making regulation almost impossible and policy complicated.

Policy for household air pollution reduction

No regulation exists for HAP anywhere in the world. Even the WHO, which has a focus in this area, has set none. The only guideline is for ambient air pollution that is often low in the village, especially when compared with the heavy loading in the cooking area. That is 25 $\mu g/m^3$.

Investigators tend to use the 25 $\mu g/m^3$ target in the evaluation of HAP interventions. Also, with the increasing use of personalized monitors, a distinction is being drawn between the exposure in the room versus that directly experienced by the cook or the child in close proximity. A United States Environmental Protection Agency (EPA) evaluation of small-scale monitors resulted in the identification of the RTI MicroPEM (Chapter 6) as having the most suitable combination of attributes. Recently, the MicroPEM has been further miniaturized to create the Enhanced Children's MicroPEM. Despite the name implying use on children, most studies use it on the cook and estimate the exposure of the child by other methods, including tracking the activity of the child.[17] The device weighs 150 g and can be worn with a strap or in a specially designed receptacle on the garment. It has an impactor that separates the $PM_{2.5}$ fraction and measures real-time $PM_{2.5}$ with a nephelometer. The particles are also captured on a Teflon filter for further in vitro study. Another feature is a six-axis accelerometer with a continuous reading.

This allows monitoring of compliance with the protocol (the instrument recognizes if the device is set aside and not worn).

The Enhanced Children's MicroPEM is the principal exposure device used in the ongoing Household Air Pollution Intervention Network (HAPIN) randomized control trial involving 3200 households in four countries: Peru, Guatemala, India, and Rwanda. This massive and expensive trial is testimony to the belief in the air quality research community that modest reductions in exposure with improved cookstoves and alternate fuels such as coke are not likely to make sufficient gains in health outcomes. This trial investigates the value of a shift to LPG as the fuel. Some countries, such as India, have instituted large-scale introduction of this fuel, but studies are lacking to justify the cost. The HAPIN trial is intended to fill that gap and to inform the details of a policy more fully. Policy in HAP could have some of the elements of the instruments described in Table 14.2.

The policy elements are informed by studies already discussed in this chapter and Chapter 8. Assurance of timely cylinder swaps is driven by the fact that the households cannot afford spare cylinders, and the distributors are not incentivized to provide prompt service to villages, especially those distant from distribution centers.[16] The need for fuel subsidies is evident in many reports, including the study in Burkina Faso, where the authors deduced an annual expenditure of US Dollar (USD) 44 per household for LPG, against an annual income of USD 162.[18] LPG pricing can be volatile. It is generally correlated with the price of oil, although in producing nations such as the United States, temporary drops in price can be created by

Table 14.2 Suggested elements of policy instruments to maximize the health benefits from substitution of cleaner burning fuel for traditional fuels such as wood and dung.

Policy instruments	Examples and details
Government-funded programs	Tax and other incentives to enablers
	Distribution of appliances to homes
	Tailoring to individual communities
	Assurance of timely cylinder swaps
Subsidy programs	Subsidies for fuel
	Education of homeowner on process
	Direct payment avoiding middlemen[a]
Regulations and policy	Standards for alternative stoves
	Exposure or outcome studies to inform policy
	Guidelines on exposure limits
	Programs for monitoring compliance

[a]Example: Aadhar card based direct payment to bank accounts (India).

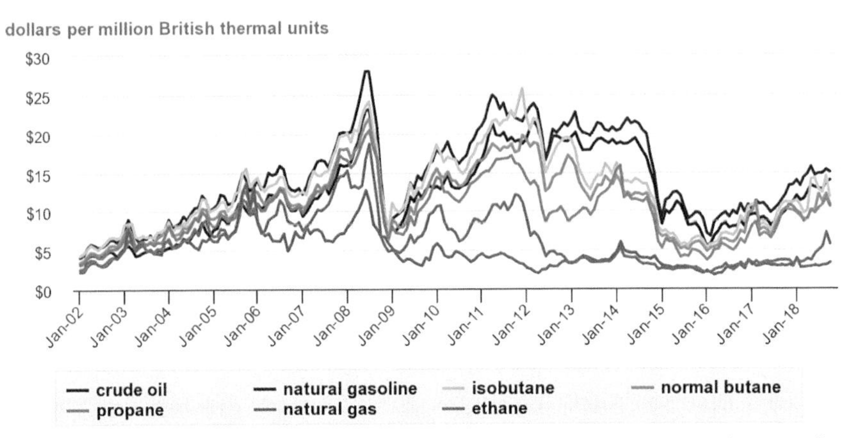

Fig. 14.3 Spot prices in USD for hydrocarbon gas liquids, oil, and natural gas in the United States. Prices are monthly averages of daily closings. Oil is Brent and natural gas is Henry Hub. *(Courtesy United States Environmental Protection Agency.)*

inability to transport to market. Most of the nations using wood as a cooking fuel are net importers of LPG, meaning they may produce some but import the rest. Because the marginal kilogram is imported, the price is set by the import price. Fig. 14.3 plots the price of hydrocarbon liquids, including propane, in the United States. Propane is the principal constituent of LPG, the balance being larger molecules, mostly butane. Consequently, propane pricing is a good proxy for LPG.

Note the tracking of the principal natural gas liquids with the price of oil. Also note the period in 2012–13 when a natural gas liquid price separation from the oil trend took place, especially for propane. This is the sort of uncertainty, coupled with the usual oil price gyrations, that makes LPG a dubious fuel for use by the poor. The price of propane was as low as USD 3.63 per million British Thermal Units (MMBTU) in January 2016 and was USD 15 in January 2014, a scant 2 years earlier. Most countries will be forced to adjust the subsidy to damp the effect on the consumer. Absent that, and perhaps even with that, the practice of stacking will proliferate the continued use of wood. Stacking could destroy most of the value of the fuel substitution.[19] A solution to just the stacking related loss of benefits could be through policy that minimizes the practice. This could include incentives to distributors to make more timely deliveries. Direct payment of subsidies into the bank accounts of households has the obvious advantages of middlemen not skimming, as happened in India with kerosene subsidies. But in communities with low empowerment of women, the men in the household could well spend the money on other things, even including alcohol. All this points to the need for policies to be community specific, which requires knowledgeable boots on the ground to advise.

Another solution is a domestically produced clean fuel. In Chapter 8, we suggest methanol as such a fuel and expect that the cost to produce would be stable and lower than that of LPG. The PM emissions should also be superior because the molecule, having no C—C bond, will produce no soot.

Policy in the United States is driven by mortality estimates because, according to the EPA, 90% of the economic value destroyed is attributable to mortality.[20] On the other hand, according to the WHO in the case of HAP, the time savings due to less time spent on wood collection and the cooking productivity gains have a value that exceeds that from averted mortality in adults and infants.[21] Therefore mortality estimates alone are insufficient to inform regulatory policy. The opportunity cost for women may differ based on the location. Policy is likely to be fine-tuned in accordance.

One final word on the HAPIN trials[22] that are underway, with no reporting of results likely till 2021. There is a definite possibility that no material gains in health outcome will be found. Preliminary data from a much smaller trial in progress of LPG use indicate reduction in exposures to below the WHO's guidelines is producing no conclusive health gains. This would be expected to a degree from an examination of Fig. 14.2. However, one of the metrics in the HAPIN study is birth weight (pregnant women are in the study group), and there could be some success here. The researchers would do well to distinguish between premature and full-term babies in the statistics. Preemies may fare differently in the test and control groups during the neonatal period than would the babies in the full-term cohort. There are hints in the literature that preemies in the two groups have less birth weight differences than in the full-term cohort.

Short-term effects on health and behavior

Most studies of particulate emissions on health outcomes cover long-term effects. However, short-term exposure is found to have affected the severity of the disease contracted from the severe acute respiratory syndrome (SARS) virus,[23] as described in Chapter 1. Whereas the correlation coefficients were not high, the trend was unmistakable, and was cited as supportive of findings in a paper on the impact of long-term $PM_{2.5}$ exposure on the severity of Covid-19.[24]

A very interesting study, apparently the first of its kind, used a metabolomics approach to evaluate the effects of $PM_{2.5}$ on adverse cardiovascular and metabolic impacts in human subjects.[25] Fifty-seven dormitory residents in Shanghai, China were subjects in a 9-day randomized, double-blind crossover study. Each dormitory room had an air purifier, half of which were sham. $PM_{2.5}$ was measured every 10 s for 24-h periods by an RTI MicroPEM (Chapter 6). Because the MicroPEM had no readout, the subjects did not know whether the purifier was sham or active. Concentrations in the sham and active rooms were on average 46.8 and 8.6 μg/m^3. The exterior concentration was 101 μg/m^3. The time weighted average of exposure for participants in sham and active rooms was 53.1 and 24.3 μg/m^3, respectively. The latter was slightly below the WHO's guideline of 25 μg/m^3 for a 24-h exposure.

The sham condition showed elevated serum glucocorticoids. For each 10 μg/m^3 increase in $PM_{2.5}$, cortisol and cortisone were elevated by 7.79% and 3.76%, respectively. Epinephrine and norepinephrine were elevated by 5.37% and 11.70%, respectively. Other metabolites were also measured, but we key here on the glucocorticoids in part to offer support for behavioral changes in response to elevated PM, which is the subject of another interesting study in London. Another takeaway from the Shanghai study is that whereas air purifiers do a very good job of reducing indoor exposures to below 10 μg/m^3, if ambient levels are high, as they were in the study, the net effect is probably not enough to avoid long-term damage.

The effect of elevated cortisol on human behavior was studied by Riis-Vestergaard et al. They conducted a double-blind study in healthy males who were injected with 10 mg of hydrocortisone (cortisol) or a placebo. This was done either 15 or 150 min prior to being required to make a decision. The decision comprised choosing between a small reward in the short term as opposed to a larger reward in the future. This is a classic behavioral economics experiment to assess discounting of future benefits. This discounting is not dissimilar to financial discounting of future profits to account for uncertainties and risk in a business. Decisions made shortly after administration of the cortisol were strongly skewed to the short term. This was not the case after a time lapse of 150 min. This suggests that corticosteroids may cause short-term temporal discounting of future rewards.[26]

Carrying that thread is a study at the London School of Economics that suggests that elevated air pollution causes increased discounting of the risk of future pain (as opposed to gain in the discussed study). The observation is that crime in London goes up on the most polluted days, pollution estimated using the air quality index (AQI). On days with AQI of 103 versus the least polluted days with AQI 9.3, the crime rate was up 8.4%.[27] They note that all these pollution levels are moderate, AQI of 100 considered by the US EPA to be "acceptable" and below 50 "good."

The same authors studied the pattern in the purchase of lottery tickets on high pollution days, noting decreased purchases that connote reduced risk taking. This appears to go in the other direction. A detailed analysis led them to conclude that, similar to the Riis-Vestergaard's study, future pain was discounted. The pain in the Bondy et al.'s study is future punishment in the form of a sentence, were the criminals caught. They reconciled the lottery purchase behavior as well, attributing the reduced purchases to the discounting of a future gain. Interestingly, major crimes such as murder and armed robbery were not increased.

Whereas at first blush all the foregoing may appear to be something of a stretch, if stress hormones are indeed elevated by high PM counts,[28] behavioral changes in response are believable. This adds an element of community well-being to the ledger of the impact of air pollution.

Regulating ultrafine particulate matter

Throughout this book, we have presented evidence for ultrafine particles (UFPs) contributing to human morbidity and mortality. The evidence has been not only health outcome based but also supported by molecular mechanisms such as penetration into lung alveolar cells being likely only by the smallest particles.[25] They penetrate membranes such as the blood-brain and placental barriers, causing memory loss and premature births, respectively.[29] A feature of all UFPs is the high surface-area-to-volume ratio that causes the surfaces to be very reactive. This reactivity causes the adsorption of organic species more readily than on larger particles. Organics are believed to be responsible for many of the adverse health outcomes.

Despite strong evidence for the toxicity of UFPs, setting regulations faces several hurdles. The chief among them is that UFPs have negligible mass, so a mass-based standard such as the current one for $PM_{2.5}$ is not feasible. In fact, the imperative for a standard is just that: the UFP mass component of $PM_{2.5}$ being negligible, massive reductions in mass per unit volume may still leave toxic UFPs behind. This could explain the current dogma that no level of $PM_{2.5}$ is safe.

Because estimating the mass of UFPs is not feasible, that leaves particle count as the measure. UFPs comprise about 80% of the particle count in ambient air,[30] so particle number concentration would be the most appropriate metric. The metric is particle number count (PNC), and measuring devices are commonly available. However, PNC may not be an appropriate metric for larger particles. A two-parameter set of regulations appears to be in our future.

Any regulatory framework requires a risk or exposure model, with an identified exposure below which the risk is considered tolerable. To inform this, one needs multiple trials targeting specific disease outcomes. A literature search of exposure or disease outcomes using PNC as the metric yielded 34 studies.[31] All were short term and used different protocols for PNC measurement, but all considered PNC to be a proxy for UFPs. All but four studies found measurable responses. However, no discernible pattern was evident for the exposure or disease function. Absent a pattern, no regulatory body would be able to provide guidance on a safe level.

Mortality estimates have always been strongly informed by epidemiology. This will be difficult in the case of UFPs because the exposure cannot be singled out to be in the ultrafine range. In vitro studies could help in this regard. Then, the question of practical relevance would remain. A first step could well have to be a clear indication that the UFP component of $PM_{2.5}$ is disproportionately toxic.

Chapter 7 describes the most physiologically relevant means for assessing the toxicity of airborne species, the CelTox method. The PM is electrostatically charged and driven down into contact with live lung epithelial cells (or any other cells, for that matter) with a drop in potential. That chapter also underlines the shortcomings of the resuspension method, the current standard. Resuspension is believed to cause agglomeration of UFPs, leading to larger particles that may not have the same biochemical action as the small ones.

A reasonably definitive study would be one, where a $PM_{2.5}$ stream from a combustion event was split so that the UFP fraction could be assessed separately. The original $PM_{2.5}$ cut would also be assessed, leading to a direct estimation of the relative toxicity of the UFPs in human cells. Then, the study could be escalated to compare effects of the two streams in animal models.

The in vitro studies could be expanded to different levels of exposure, likely estimated by PNC. Human studies may still be difficult. Air purifiers such as those used in the Li et al.'s study bring the loading down to just 8 $\mu g/m^3$. At that level, most of the UFPs are almost certainly still in play. This feature could be used in an experimental design.

Regulation of UFPs will almost certainly use the PNC metric. Even though UFPs are a subset of $PM_{2.5}$, and even if they are demonstrably the more toxic component of the fraction, mass-based regulation cannot be replaced with PNCs. In part this is because the larger particle count is a fraction of that of UFPs, and the larger species up to 10 μm independently have health or visibility consequences.

Close-up: Taking the lead out

As a nearly freshly minted PhD, I (V.R.) took the Exit 8 off-ramp of the New Jersey Turnpike. It was marked by a water tower with a Dutch Boy Paint logo. I did not know then that Dutch Boy Paint was a part of NL Industries, the company I was going to join as a researcher. In looking them up, I discovered that 2 years prior to this February 1974 date, they had taken the lead out of their name. The previous name had been National Lead Company. I was joining a group that researched recycling several metals, including aluminum, zinc, and lead. The lead part was still there, so that could not be the reason for the name change. I realized then that it was just the name association with a toxic substance. Lead was coming out, not just out of goods but also out of names.

Lead was indeed coming out of goods. More prominently, lead was taken out of gasoline and indoor paint. The use of oxides of lead, known as white lead, goes back to artists' paint. Color could be added to the basic white, and lead oxide was believed to provide the most lasting color. The toxicity of lead was most injurious to children, especially not only from the paint on toys but also from interior paint. The removal took decades of scuffling and culminated in a ban on lead in indoor paint in 1978. At least three decades prior to the ban, titanium dioxide was known to be a viable substitute. In fact, it was whiter than white lead and is today used in most things one sees white, including toothpaste and paper. But, in an ode to the axiom that there are no free lunches, the extraction of TiO_2 from the ore ilmenite, $FeO{\cdot}TiO_2$, involved the use of sulfuric acid and the disposal of the ferrous sulfate in a form known copperas. Although benign (finding use as an Fe supplement in agriculture), the residual acid proved to be a problem for some disposal schemes.

Lead in gasoline was invented in the early 1920s. Tetraethyl lead (TEL) was found to increase the octane rating of gasoline with very small admixtures when compared with the only other such agent known at the time, ethanol. (As fate would have it, a 100 years later ethanol is back.) The increase in engine performance was substantial because of the high compression ratios now possible (Chapter 11), and TEL in gasoline became a standard. However, engine exhaust emissions had enough lead to be a serious health hazard. Several decades on, what tipped the scales on discontinuance of TEL was that the lead was fouling the newly invented three-way catalytic converter. When other octane enhancers replaced TEL by law, blood lead levels in the population plummeted in lock step. This was one of the more dramatic examples of the efficacy of a regulation. Curiously, Thomas Midgley, the inventor of TEL, also invented chlorofluorocarbons. This transformed refrigeration. But it also proved to be ozone depleting in the atmosphere (Chapter 3). Chlorofluorocarbons were banned from use in compressors in 1996, the same year that the full ban on lead in gasoline was implemented. Thomas Midgley had the extraordinarily good luck to make two industry-changing discoveries, and the equally extraordinarily bad luck to have both of his discoveries regulated out of existence.

Now, a return to memory lane. Lead was coming out of shotgun pellets. NL was a major manufacturer of lead shot, the pellets that filled the shell. The pellets that did not hit a bird (no doubt the vast majority for the average hunter of fowl) could fall into water bodies. Ducks grind their food using pebbles. Apparently, they pick up pebbles by feel rather than sight. In any case lead pellets get into the grinder mix and lead poisoning ensues. Regulators banned the use of lead shot in the flyways. Steel shot was required. The density of lead causes the shot to stay in a tight pattern. Steel patterns are not tight. Hunters hated steel shot.

The rookie researcher to the rescue. I invented a lead shot with a small addition of a substance that caused the shot to fall apart when submerged in water for a period. I figured that metallic lead pieces would not leach in the water and ducks would not be misled (sorry, had to do that). Fortunately, I bragged on this to a guy who sold shot to reloaders before I informed supervision. He professed to be impressed. Then this.

Joe: "So, are you a hunter of fowl?"

Me: "Never held a gun in my life."

"Figured that. What do you suppose will happen when the duck is cooked?" says Joe, with this gotcha expression on his face. I am on guard now. Even I knew that hunters did not attempt to remove shot from the bird before it suffered other indignities prior to being cooked.

"Oh, (expletive)," I said. The lead pellets, ordinarily discernible by sight or feel, would disintegrate in the cooking juices and acids and heat. Nothing good would result after that in the diner's body. I had just invented a product that would reduce the population of hunters. Worse, these would be customers. I swore that day to never again attempt to invent anything without first vetting the concept of said invention with the folks in the field.

References

1. World Health Organization. *Ambient Air Pollution: A Global Assessment of Exposure and Burden of Disease*; 2016. ISBN:9789241511353.
2. Pope III CA, Cropper M, Coggins J, Cohen A. Health benefits of air pollution abatement policy: role of the shape of the concentration–response function. *J Air Waste Manage Assoc.* 2015;65(5):516–522. https://doi.org/10.1080/10962247.2014.993004.
3. Burnett RT, Pope III CA, Ezzati M, et al. An integrated risk function for estimating the global burden of disease attributable to ambient fine particulate matter exposure. *Environ Health Perspect.* 2014;122:397–403. https://doi.org/10.1289/ehp.1307049.
4. Chartier R, Phillips M, Mosquin P, et al. A comparative study of human exposures to household air pollution from commonly used cookstoves in Sri Lanka. *Indoor Air.* 2017;27(1):147–159. https://doi.org/10.1111/ina.12281. 26797964.
5. Genualdi SA, et al. Trans-Pacific and regional atmospheric transport of polycyclic aromatic hydrocarbons and pesticides in biomass burning emissions to western North America. *Environ Sci Technol.* 2009;43(4):1061–1066.
6. Killin RK, Simonich SL, Jaffe DA, DeForest CL, Wilson GR. Transpacific and regional atmospheric transport of anthropogenic semivolatile organic compounds to Cheeka Peak Observatory during the spring of 2002. *J Geophys Res Atmos.* 2004;109(D23), D23S15.
7. Rogers HM, Gitto JC, Gentner DR. Evidence for impacts on surface-level air quality. *Atmos Chem Phys.* 2020;20:671–682. https://doi.org/10.5194/acp-20-671-2020.
8. Lin C-Y, Liu SC, Chou CC-K, et al. Long-range transport of aerosols and their impact on the air quality of Taiwan. *Atmos Environ.* 2005;39(33):6066–6076.
9. Giannadaki, Lelieveld J, Pozzer A. Implementing the US air quality standard for $PM_{2.5}$ worldwide can prevent millions of premature deaths per year. *Environ Health.* 2016;15:88. https://doi.org/10.1186/s12940-016-0170-8.
10. Zhang Y, West JJ, Mathur R, et al. Long-term trends in the ambient $PM_{2.5}$ and O_3-related mortality burdens in the United States under emission reductions from 1990 to 2010. *Atmos Chem Phys.* 2018;18:15003–15016. https://doi.org/10.5194/acp-18-15003-2018.
11. Fann N, Kim SY, Olives C, Sheppard L. Estimated changes in life expectancy and adult mortality resulting from declining $PM_{2.5}$ exposures in the contiguous United States: 1980–2010. *Environ Health Perspect.* 2017;125:2017. https://doi.org/10.1289/EHP507.
12. Pope III CA, Burnett RT, Turner MC, et al. Lung cancer and cardiovascular disease mortality associated with particulate matter exposure from ambient air pollution and cigarette smoke: shape of the exposure–response relationships. *Environ Health Perspect.* 2011;119:1616–1621. https://doi.org/10.1289/ehp.1103639.
13. Malik K. *Essays on Energy and Environment in India* [Ph.D. dissertation]. College Park, MD: University of Maryland; 2013.
14. Gould C, Urpelainen J. LPG as a clean cooking fuel: adoption, use and impact in rural India. *Energy Policy.* 2018;122:395–408.
15. Srinivas M. LPG Consumers, Stop Being Short-Changed. *The Hindu.* 2016. updated May 28, 2016 https://www.thehindu.com/news/cities/Hyderabad/lpg-consumers-stop-being-shortchanged/article5339986.ece. Retrieved 3 June 2020.
16. Sehgal Foundation. *A Study on Variation in Perceptions and Preferences for Cooking Fuel Choice and Cooking Methodologies in Samastipur District of Bihar, India*; 2020. report awaiting publication (private communication).
17. Johnson M, Steenland K, Piedrahita R, et al. Air pollutant exposure and stove use assessment methods for the Household Air Pollution Intervention Network (HAPIN) trial. *Environ Health Perspect.* 2020;128(4). https://doi.org/10.1289/EHP6422.

18. Yamamoto S, Sié A, Sauerborn R. Cooking fuels and the push for cleaner alternatives: a case study from Burkina Faso. *Glob Health Action*. 2009;2:1–9. https://doi.org/10.3402/gha.v2i0.2088.
19. Johnson MA, Chiang RA. Quantitative guidance for stove usage and performance to achieve health and environmental targets. *Environ Health Perspect*. 2015;123:820–826. https://doi.org/10.1289/ehp.1408681.
20. United States Environmental Protection Agency. *The Benefits and Costs of the Clean Air Act from 1990 to 2020 (Final Report—Rev A)*. Washington, DC: U.S. Environmental Protection Agency Office of Air and Radiation; 2011.
21. World Health Organization. *Fuel for Life, Household Energy and Health, Booklet*; 2006.
22. Clasen T, Checkley W, Peel JL, et al. Design and rationale of the HAPIN study: a multicountry randomized controlled trial to assess the effect of liquefied petroleum gas stove and continuous fuel distribution. *Environ Health Perspect*. 2020;128(4):1–11. https://doi.org/10.1289/EHP6407.
23. Cui Y, Zhang ZF, Froines J, et al. Air pollution and case fatality of SARS in the People's Republic of China: an ecologic study. *Environ Health*. 2003;2:15.
24. Wu X, Nethery RC, Sabath BM, Braun D, Dominici F. Exposure to air pollution and COVID-19 mortality in the United States. *medRxiv*. 2020. https://doi.org/10.1101/2020.04.05.20054502.
25. Li N, Georas S, Alexis N, et al. A work group report on ultrafine particles (American Academy of Allergy, Asthma & Immunology): why ambient ultrafine and engineered nanoparticles should receive special attention for possible adverse health outcomes in human subjects. *J Allergy Clin Immunol*. 2016;138(2):386–396. https://doi.org/10.1016/j.jaci.2016.02.023.
26. Riis-Vestergaard M, van Ast V, Cornelisse S, Joëls M, Haushofer J. The effect of hydrocortisone administration on intertemporal choice. *Psychoneuroendocrinology*. 2018;88 (February):173–182. https://doi.org/10.1016/j.psyneuen.2017.10.002.
27. Bondy M, Roth S, Sager L. *Crime Is in the Air: The Contemporaneous Relationship Between Air Pollution and Crime*. IZA Institute of Labor Economics; April 2018. Discussion paper series, IZA DP No. 11492.
28. Li H, Cai J, Chen R, et al. Particulate matter exposure and stress hormone levels. *Circulation*. 2017;136(7):618–627. https://doi.org/10.1161/CIRCULATIONAHA.116.026796.
29. Kang D, Kim J-E. Fine, ultrafine, and yellow dust: emerging health problems in Korea. *J Korean Med Sci*. 2014;29(5):621–622. https://doi.org/10.3346/jkms.2014.29.5.621.
30. Kumar P, Robins A, Vardoulakis S, Britter R. A review of the characteristics of nanoparticles in the urban atmosphere and the prospects for developing regulatory controls. *Atmos Environ*. 2010;44(39):5035–5052.
31. Zelasky S. *Review of Evidence Supporting a New Count-Based Particulate Matter Standard in the US. Air Pollution Course Final Paper*. Harvard T. H. Chan School of Public Health; May 2020.

Index

Note: Page numbers followed by *f* indicate figures, and *b* indicate boxes.

E

F

M

N

O

P

Printed in the United States
by Baker & Taylor Publisher Services